HAWKING

Published in 2018 by André Deutsch
An imprint of the Carlton Publishing Group
20 Mortimer Street
London W1T 3JW

10 9 8 7 6 5 4 3 2 1

Text © Joel Levy 2018
Design © André Deutsch 2018

The right of Joel Levy to be identified as the author of
this work has been asserted by him in accordance with
the Copyright, Designs and Patents Act 1988

A CIP catalogue record for this book is available from
the British Library.

ISBN 978 0 233 00570 6

Printed in Italy

HAWKING

THE MAN, THE GENIUS AND THE THEORY OF EVERYTHING

ANDRE
DEUTSCH

CONTENTS

INTRODUCTION

The death of Stephen Hawking in March 2018 brought to a close one of the most incredible narratives of the last century, a story that included scientific discoveries at the frontiers of space and time, romance against all the odds and heroic willpower in the face of crushing adversity. If this tale sounds as if it would make a good movie – don't worry, it's already been made into two films, so far…

Stephen Hawking's life captured the imagination and admiration of millions around the world. He created a publishing phenomenon, met with popes and presidents, and filled concert halls in the manner of a rock star. He travelled the world, experienced zero gravity and hot air balloon flights, guest-starred in the world's most popular television series and was portrayed on the silver screen by movie stars. Hawking won an extraordinary slew of awards, accolades and honours, ranging from Britain's Order of the Companions of Honour and the US Presidential Medal of Freedom to the Albert Einstein Medal, the Aventis Book Prize and the Pius XI Medal from the Pontifical Academy of Sciences.

After the runaway success of his 1988 *A Brief History of Time,* Hawking became the iconic intellectual of his age. In an era of ever-worsening dumbing down, he presented the media and the general public with the perfect shorthand for complex and challenging concepts, such as science and genius. His public image transcended human standards to the realm of archetype: a wasted body and computer-synthesized voice, just a few steps removed from a brain in a vat; pure intellect freed of gross corpority. But to let this presentation, constructed and imposed from outside Hawking's own life, obscure the reality of his scientific and personal achievement, is a disservice and a failure of the imagination. His refusal to let his illness define him, let alone stop him, is inarguably inspiring. While he may not have been the greatest cosmologist since Einstein or even in the top rank of modern physicists, the topics he explored in his life's work are fascinating and thrilling.

This book seeks to show that anyone can join in, to some extent, the intellectual adventure on which Hawking embarked. It explains the background to his science, and tries to make accessible the intricacies of his theories, as far as is possible without mathematics. Hawking recalled being warned that every equation he included in a book would halve his readership; I have followed his dictum and avoided almost all equations, with the exception of the one he chose as his epitaph.

CHAPTER 1

LAZY EINSTEIN

THE REPLACEMENT GALILEO: FAMILY AND EARLY YEARS

STEPHEN WILLIAM HAWKING WAS BORN ON 8 JANUARY 1942; IT WAS 300 YEARS TO THE DAY SINCE THE DEATH OF GALILEO, AS STEPHEN LATER ENJOYED POINTING OUT. HIS PARENTS, FRANK AND ISOBEL, HAD BOTH ATTENDED OXFORD UNIVERSITY, DESPITE COMING FROM FAMILIES THAT WERE NOT WEALTHY.

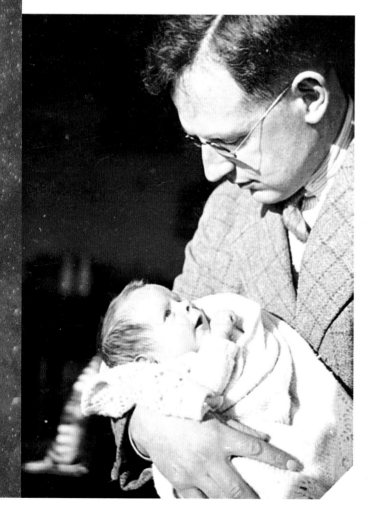

Isobel was the daughter of a Scottish doctor, and her family scraped together the funds to put her through a university education at a time when this was extremely unusual for a woman. Stephen's father, Frank, came from the north of England, from Yorkshire (famed for its plain-speaking inhabitants); he studied tropical medicine, going on to become a leading researcher whose work often took him overseas.

Stephen's parents met while both were working at a medical research institute in north London, and married in the early years of the Second World War. The family home was in Highgate in north London, but the danger of German bombing led Isobel to relocate to Oxford while she was in the late stages of pregnancy with Stephen. Visiting a bookshop a few days before giving birth, she bought an astronomical atlas, a purchase she later regarded as a sign of things to come. On 8 January 1942, Isobel gave birth to her first child – Stephen William Hawking.

Left: Frank Hawking holding his infant son Stephen William in 1942. At this time the Hawkings were living in north London.

Right: Stephen as a young boy, playing around in a boat. Though already known for being physically awkward, his illness had not yet manifested.

After Stephen's birth Isobel returned to Highgate, where the family would live for the next eight years, and went on to have two more children (Mary, born 1943, and Philippa, born 1946); later, in 1955, she and her husband would adopt a fourth child, Edward. While in Highgate, Stephen attended Byron House, a school with a progressive educational philosophy that he blamed for his difficulties in learning to read. In 1950 Frank was appointed head of the Division of Parasitology at the National Institute for Medical Research in St Albans, a prosperous provincial town to the north of London, and the family relocated there.

In St Albans the Hawkings became known for being rather eccentric. The family car was a repurposed London taxicab, and the family home was a large, ramshackle house that started off dilapidated and became more so, as repairs were rarely attempted. Towers of books helped cover up some of the cracks and provided a little insulation against the draughts. Frank Hawking himself was not much concerned with such domestic issues, as he was often absent for long periods, travelling in tropical countries. At home, the Hawkings were known to sit around the dining table in silence, each lost in a book, and they were also renowned for their rapid-fire mode of speaking (*see* box).

HAWKINGESE

Frank Hawking spoke with a stutter and the rest of his family were notorious for gabbling. Stephen's friends had a theory that the family were so clever that their thoughts tumbled out too fast for their mouths to keep up. One result was that Stephen and his family often mangled and contracted words and phrases to produce a peculiar sort of garbled English that friends labelled "Hawkingese".

AN HOUR A DAY: SCHOOL AND UNIVERSITY

AFTER THE HAWKING FAMILY MOVED TO ST ALBANS, THE EIGHT-YEAR-OLD STEPHEN BRIEFLY ATTENDED THE HIGH SCHOOL FOR GIRLS (WHICH WAS ACTUALLY CO-EDUCATIONAL, DESPITE ITS NAME), WHERE HIS DISTINCTIVE MOP OF HAIR WAS NOTICED BY A GIRL IN THE NEXT CLASS: SEVEN-YEAR-OLD JANE WILDE, WHO WOULD LATER BECOME HIS WIFE.

Above: Schoolboy Stephen Hawking. Like some other notable scientists, Hawking did not initially excel at school, mainly through lack of interest in formal education.

Eventually Stephen moved on to St Albans School, where he pursued a curious school career; although clearly highly intelligent, he failed to apply himself and was usually far from top of the class. His classmates called him "Lazy Einstein".

However, Stephen read widely, educating himself and developing a fascination for science. Like many children, he constantly asked questions about the causes of natural phenomena, but unlike many adults, this childlike questioning nature never left him, and Stephen would later identify it as a driving force behind his career. He also began to display signs of the intellectual arrogance that would become a characteristic feature; his school friend Michael Church, for example, recalled Stephen making a fool of him in a philosophy debate.

A notable exploit of Stephen's school years was his role in the construction of an early electronic computer as a school project. Known as LUCE (Logical Uniselector Computing Engine), this device was made from an assortment of oddments including old clocks and a recycled telephone switchboard. In 1957/8 Stephen and his friends worked on the project, which even made it into the local papers. Forty years later Hawking would play a pivotal role in the creation of COSMOS, a cosmological supercomputer set up to crunch vast quantities of data and dedicated to research in cosmology, astrophysics and particle physics.

After school, Stephen's father was keen for him to pursue medicine; but Stephen's interests lay in the purer sciences. His father did,

however, prevail in his choice of college: in October 1959 Stephen "went up" to his father's alma mater, University College, Oxford. Yet even there he resisted full engagement with education. At that time it was considered uncool to work very hard, at the risk of being considered a "grey man". Stephen took this very much to heart, doing as little work as he could possibly get away with. He later calculated that in three years at Oxford he did only around a thousand

IS IT POSSIBLE TO TELL IF SOMEONE WILL BE A GENIUS?

A comparison of the early biographies of Isaac Newton, Albert Einstein and Stephen Hawking reveals many similarities. None of the three excelled at school, at least initially, perhaps because their intellects were not fully engaged; for instance, as with Hawking, Einstein's schooling was marked by laziness and insolence. All three were auto-didactic and often dreamy or distracted, although when motivated they could quickly get to the top of the class. Stephen, for example, despite years of mediocre performance at school, secured a scholarship to Oxford a year earlier than normal. Additionally, as children all three scientists delighted in making models or tinkering with mechanical toys or gadgets.

Right: Einstein, the German-born physicist famed for developing the theory of relativity.

hours of work in total, equating to just one hour a day. He spent far more time being an over-adventurous coxswain for the University College Boat Club than engaged in his studies.

It was while he was at Oxford that Stephen's increasing physical unsteadiness began to manifest itself. He had always been gawky and gangling, but now he became dangerously

Above: The Oxford University Boat Club in 1962. Hawking is posing distinctively to the right of the picture, holding aloft a handkerchief. His friend Gordon Berry is upside down in the middle.

Below: A Cambridge newspaper records the result of a disappointing showing by the Oxford coxswains' boat crew (including Stephen) in a race against Cambridge. At this time Hawking seemed more interested in rowing than in his studies.

Light Blues Win This Boat Race

THE annual coxswains' Boat Race between Oxford and Cambridge on the Cam on Saturday resulted in a 15-lengths victory for the Light Blues.

In very good conditions, the race, umpired by Mr. P. J. D. Allen, the captain of Jesus College Boat Club, was rowed from Ditton Corner to Peter's Posts. Both crews got off to a good start, Cambridge striking 35, Oxford 32. Cambridge established a length's lead in the first ten strokes, which they gradually increased.

The standard of rowing of both crews was a great deal higher than in previous years, which reflects great credit on the coaches, M. Reupke (Jesus, Cambridge) and J. L. Stretton (Trinity, Oxford), since neither crew had had more than the statutory six outings.

This race brings the score to ten victories for Cambridge and two for Oxford, since the event was started in 1949.

The Cambridge crew, which is so far unbeaten, has issued a challenge to "any irregular combination of lighthearted oarsmen, to a race from Ditton to Peter's Posts (or vice versa, depending on the direction of the wind) for a stake of nine pints" If the challenge is accepted, the racing will be at 2 p.m. on Thursday, March 10th.

C.U.C.S.—P. Gaskell-Taylor (Peterhouse), bow; J. A. M. Butters (Lady Margaret B.C.); D. M. Gill (Corpus); R. K. B. Hankinson (Trinity Hall); T. J. Hannah (Jesus), J. R. W. Keates (First and Third); S. Lomas (Downing); J. A. Morrison (Jesus), stroke; J. A. D. Innes (Fitz.), cox.

O.U.C.S.—H. G. Berry (Univ.), bow; M. P. Connon (B.N.C.); B. R. Wilson (Hertford); A. A. Jones (Lincoln); P. Groves (St. Cath's Soc.); S. W. Hawkong (Univ.); R. D. Finlay (Christ Church); M. H. Griffiths (Oriel), stroke; D. Hilton (Hertford), cox.

Printed by "Cambridge Daily News" (1929) Limited. A. C. Taylor, Managing Director. Theatre Buildings, St. Andrew's Street, Cambridge.

clumsy. On one occasion he fell downstairs at the college and bashed his head quite badly, suffering from temporary memory loss. His friend Gordon Berry recalls that, in order to make sure he had not sustained any major damage, Stephen took an IQ test, achieving an astronomically high score.

When it came to his finals, Stephen tried to cover for his failure to apply himself to his education (especially the practical aspects) by answering only questions on theoretical physics. According to his version of events, his results put him on the borderline between getting a first and second class degree. When interviewed by a panel to decide which degree to award, he told them that he planned to pursue academic research, and that if he got a first he would go to Cambridge, while if he got a second he would stay at Oxford. Keen to get rid of this troublemaker, the panel duly awarded him a first. Stephen was about to take his first steps in a scientific career that would shake the foundations of physics.

Above: Stephen and friends making their way back to college from the University College boat house in 1961. Hawking is on the left in a boater.

Opposite above: Hawking (third from left, seated in an armchair) and members of the rowing crew he had coxed to victory in a regatta in 1960.

ON ANOTHER PLANET

Despite his lack of application, Hawking's extraordinary intelligence was evident to tutors and students alike. A friend of his from Oxford, Derek Powney, who was one of the other physics students in his class, recalled a revealing anecdote from their second year of studies. Four students had been set a list of 13 extremely tough physics questions for homework, and the other three students, despite labouring for a week, managed to complete only two of them. Hawking didn't bother to do any. On the morning of their tutorial the others forced a reluctant Hawking to get up for breakfast for a change, although he declined to attend lectures in the morning. When his three classmates got back to college three hours later they jokingly asked him how many questions he had done. "I've only had time to do the first ten," he replied. It was at this point they realized that, when it came to intelligence, Hawking "was on another planet".

Right: Hawking (right) with his friends Gordon Berry (left) and Derek Powney (centre).

CHAPTER 2

A DEATH SENTENCE AND A REPRIEVE

CHOOSING COSMOLOGY: GOING TO CAMBRIDGE AND STARTING A PHD

IT WAS IN HIS THIRD YEAR AT OXFORD THAT STEPHEN, FACED WITH A CHOICE BETWEEN SPECIALIZING IN COSMOLOGY OR ELEMENTARY PHYSICS, STARTED TO THINK BIG.

He chose to focus on cosmology, the study of the universe and the physics behind topics such as the formation of stars and the evolution of galaxies, as well as more exotic phenomena such as black holes. Cosmology is also about perhaps the biggest question in science: where did the universe come from?

Hawking took a summer course with Jayant Narlikar, a graduate student of Fred Hoyle, the pre-eminent British cosmologist of the age. The course kindled in Hawking a desire to do a PhD at Cambridge under Hoyle. Having secured his first from Oxford, Hawking arrived at Trinity Hall, Cambridge,

somewhat the worse for wear. Increasing anxiety over his movement problems had not been ameliorated by a stressful summer holiday in Persia (present-day Iran), where he fell dangerously ill at the same time as being caught up in an earthquake.

In academic terms, the main challenge facing him was to find a topic for his PhD research. Hawking's abiding fascination with big questions about the universe led him to an interest in cosmology in general and Einstein's theory of relativity in particular. He knew that fundamental or particle physics – the other main branch of theoretical physics that he

ASSIGNED TO SCIAMA

Matters did not improve when Stephen arrived at Cambridge. To his dismay, Hoyle already had a full quota of graduate students, and instead Stephen was assigned to a man he had never heard of, Dennis Sciama, then a lecturer in mathematics and an ally of Hoyle's cosmology and his steady state theory (see pages 22–3). In fact, this was a stroke of luck for Stephen: Hoyle travelled often and had little time for his graduate students, while Sciama was a gifted educator renowned for mentoring his students. Additionally, his research interests coincided with Hawking's hunger to explore fundamental problems of cosmic significance.

Left: Dennis Sciama, whose good sense and compassion would help pull Stephen out of his depression and start him on the road to scientific greatness.

The equations on the chalkboard:

$$R_{ik} - \frac{1}{2}g_{ik}R$$

$$= -\kappa T_{ik} + ?$$

might have explored – was of less interest to him. He felt particle physicists to be more like botanists or zoologists, doing little more than classifying a "zoo" of exotic particles without the benefit of a well-defined underlying theory.

Relativity, on the other hand, Hawking viewed as being based on a well-defined theory with equations set out by Einstein, but with great scope for exploration because the equations were so challenging that the theory and its cosmological implications had not been well explored. However, settling on relativity brought its own problems, because the mathematics were fiendishly complex and Hawking lacked the necessary grounding. Accordingly, he began to travel to London to attend a course in relativity.

Above: Indian astrophysicist Jayant Narlikar in later life. As a young doctoral and post-doctoral student he studied with Fred Hoyle and became friends with Hawking.

FROM THE ABSTRACT TO THE CONCRETE

The 1960s would prove to be an exciting time for cosmology. Until then relativity had been a topic primarily for mathematicians; it was not felt to have too much relevance to applied physics. However, the birth of radio astronomy in the 1950s began to reveal a new world of astronomical phenomena such as pulsars, and relativity would prove key to understanding them. Around the same time, physicists such as John Wheeler, building on work done by Robert Oppenheimer and others, began to articulate theories about black holes. Abstract theories from mathematical physics were gaining startling confirmation from concrete astronomical observations, electrifying the esoteric world of cosmology.

HOYLE AND THE STEADY STATE

FRED HOYLE WAS THE LEADING FIGURE IN BRITISH ASTRONOMY AND COSMOLOGY IN THE POSTWAR YEARS. A CELEBRITY SCIENTIST WHO HAD MADE BREAKTHROUGH DISCOVERIES IN THE SCIENCE OF STARS AND THE FORMATION OF ELEMENTS, HOYLE WAS PARTICULARLY WELL KNOWN AS THE LEADING EXPONENT OF THE STEADY STATE THEORY OF COSMOLOGY.

After the Second World War, Hoyle had earned his place in the textbooks by working out the theory of stellar nucleosynthesis, the process by which elements are forged (that is, their nuclei are synthesized) through fusion inside stars. This work coined the well-known phrase "we are all made of stardust". Having helped to answer one big question – "where do we come from?" – Hoyle now turned his attention to another: "how did it all begin?".

Evidence for an expanding universe, such as Hubble's redshift discoveries (*see* page 36), led many to suggest that the universe must have started off tiny and exploded into its current enormity. In 1945 Hoyle had started giving a series of radio broadcasts with a "fireside chat" format, in which he conversationally introduced the general public to profound concepts in astrophysics. In the very first of these broadcasts Hoyle had coined a dismissive term for the theory of an explosive cosmological origin: "the Big Bang".

Dissatisfied with the Big Bang theory and its implication that the universe had "magically" appeared out of nothing, Hoyle championed an alternative advanced by his friend and colleague Thomas Gold, which suggested that in fact the universe had always existed and always would. Working with Gold and Hermann Bondi, Hoyle developed a theory to

Opposite: Fred Hoyle, the leading British astrophysicist in the immediate postwar years and an influential science communicator as well as a theorist.

Right: Hoyle's nemesis, Martin Ryle, pictured in front of one of the radio telescopes that was electrifying the world of cosmology by opening new windows on the universe.

show that the universe appears to be expanding because new galaxies are constantly being created to replace those that die out. Creation and extinction are in equilibrium, producing a steady state, and so it became known as the "steady state theory".

Hoyle's nemesis was Martin Ryle, a radio astronomer who organized a massive survey of radio sources in galaxies across the universe. In Hoyle's continuous-creation model, galaxies, with their radio sources, are produced at a constant rate and should therefore be evenly distributed around the universe. However, the Big Bang model predicts that since radio sources were created near the birth of the universe, they should be old and therefore far away. In 1955 Ryle announced the results of his "Cambridge survey" of radio sources, which seemed clearly to support the Big Bang model, but his moment of triumph was spoiled when mistakes in the process came to light. "I have little hesitation in saying that a sickly pall now hangs over the Big Bang theory," crowed Hoyle, but Ryle would have the last laugh. By 1961, after repeated revisions, the fourth Cambridge survey seemed to settle the matter in Ryle's favour. Hoyle refused to admit defeat, insisting as late as 1999 that the Big Bang theory is "a huge facade based on no real evidence at all".

HOYLE THE CONTRARIAN

Hoyle was noted for his contrarian views. He retired from academia in 1972, complaining that there was too much politics involved, but continued to champion dissenting views. He supported the concept of panspermia (the theory that life on Earth was seeded by viruses arriving from space on comets), argued that Stonehenge had been built as an eclipse predictor, attacked Darwinism and supported a version of intelligent design (a creationist theory).

Right: Biomolecules scattered in space; according to the theory of panspermia, Earth was seeded with life by extraterrestrial organics.

A BIT OF A SHOCK: DIAGNOSIS WITH ALS

STEPHEN'S FIRST TERM AT CAMBRIDGE AND HIS STRUGGLES IN PICKING A PHD TOPIC WERE OVERSHADOWED BY HIS WORSENING CO-ORDINATION AND MOVEMENT PROBLEMS.

He found it increasingly difficult to perform tasks such as tying up his shoelaces, while his speech was often slurred and he was becoming ever more clumsy. When he went home to St Albans for Christmas, his father took him to the doctor and he was referred to St Bartholomew's hospital in London. In January 1963, just after his twenty-first birthday, Hawking travelled there for extensive neurological testing, including painful injections of radioactive fluid into his spine for scanning.

The medical diagnosis, when it eventually came, was appalling. Stephen had amyotrophic lateral sclerosis (ALS), a form of motor neurone disease known in the US as Lou Gehrig's disease, after the New York Yankee baseball player who died from it in 1941. Initial hopes that Stephen's condition was stable proved unfounded and it was apparent that in fact

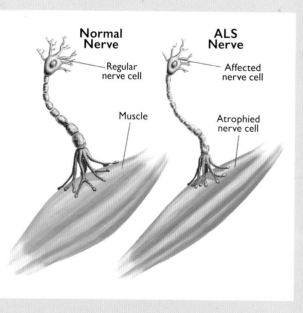

ALS

ALS is an incurable and untreatable condition in which nerve cells in the spinal cord and brain, which regulate voluntary muscular activity, degenerate, in turn causing their associated muscles to atrophy. Automatic muscles, such as the heart, are not affected, and neither is the brain, but the condition is usually profoundly life-limiting because it leads within a few years to failure of the respiratory muscles, resulting in pneumonia or suffocation. Before this happens, sufferers lose the ability to speak, eat or move, although muscles in the gut and reproductive systems are not affected.

Left: US baseball legend Lou Gehrig, who suffered from ALS, shedding a tear during a tribute at Yankee Stadium in New York, 1939.

Below: Brain scan of an ALS patient. The disease affects voluntary muscle controlling nerve cells in the brain and the rest of the nervous system.

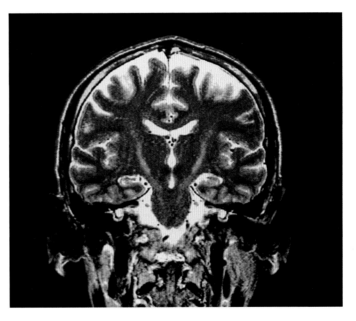

it was deteriorating. While the doctors could not know how quickly his condition would progress, they were forced to issue a truly dire prognosis – Stephen probably had only two years to live. With typical understatement, Hawking recalled, "The realization that I had an incurable disease that was likely to kill me in a few years was a bit of a shock."

Understandably, he fell into what his future wife, Jane, described as "a deep depression…. Harsh black cynicism, aided and abetted by long hours of Wagnerian opera at full volume." Holed up in his room at Cambridge, Hawking listened to music, read science fiction, struggled with terrifying nightmares and showed little interest in his PhD, which was not going well. Frank Hawking appealed to Stephen's supervisor to guide him to a thesis topic that he could complete in the short time available, but Sciama, who could see Stephen's incredible potential and his desire to tackle fundamental topics, wisely demurred.

HAWKING AND ALS

Stephen Hawking managed to live for over 50 years with an incurable degenerative disease that normally kills within two years. Doctors cannot say for sure how he achieved this feat, but it is thought to illustrate the variability of the disease between individuals. Normally ALS is fatal because it causes failure of respiratory muscles, but in Hawking's case, his manifestation of the disease preserved enough function for him to survive. In addition he had excellent, intensive care and support to help lessen and overcome the risk of lung infections.

STRANGE BUT VERY CLEVER: MEETING JANE

IN THE WAKE OF HIS SHOCKING DIAGNOSIS, THE ONE BRIGHT SPOT IN HAWKING'S LIFE WAS HIS BLOSSOMING ROMANCE WITH JANE WILDE, A GIRL WHOM HE HAD ORIGINALLY MET AT A NEW YEAR'S EVE PARTY IN ST ALBANS, THEIR HOME TOWN, AT THE END OF 1962.

He had caught her eye a few months earlier when she spied him on the other side of the street, "a young man with an awkward gait", she recalled in her 2007 memoir *Travelling to Infinity: My Life With Stephen*, "his head down,

his face shielded from the world under an unruly mass of straight brown hair". A friend had described him as "strange but very clever", and Jane found herself "drawn to this unusual character by his sense of

Above: Jane and Stephen, as played by Felicity Jones and Eddie Redmayne, at the May Ball in Cambridge, in the 2014 film adaptation of Jane's memoir.

JANE WILDE

Jane was a St Albans girl who had attended the same school as Stephen, in the year below him. When they met, she had just been accepted to Westfield College at the University of London to study languages. Her religious faith was at odds with Stephen's atheism, and she was initially both attracted and put off by his intellectual arrogance. Jane's optimism, however naïve, would prove to be a source of comfort and inspiration to Stephen, and he would later credit her spirit with helping him find the courage to go on in the face of his diagnosis. Their married life and the responsibilities that came with it meant that Jane put her own academic career on hold, but in 1981 she would gain a PhD of her own – in medieval Spanish poetry. She was also keen on music and particularly enjoyed singing in a chorus.

Right: Newlyweds Stephen and Jane. Jane deliberately avoided learning too much about Stephen's prognosis, although he was already severely affected.

humour and independent personality". This was before Stephen's diagnosis, which Jane heard about when she was studying in London. A week later she bumped into him on the platform at the train station. He invited her to a ball at Cambridge and she went to the opera with him in London, but he barely acknowledged his illness, and she found herself having to cope with his falls and the way that "as his gait became more unsteady, so his opinions became more forceful and defiant". These were foreshadows of their life together to come.

Bitterness about Stephen's illness threatened to end his budding relationship with Jane.

Playing croquet on the lawn of his Cambridge college, he set out to humiliate her, "scarcely bother[ing] to veil his hostility and frustration, as if he were deliberately trying to deter me from further association with him". Yet, she later confessed, Jane was already under his spell, and Stephen seemed to realize the value of their bond after they had spent the summer apart. That October he whispered a marriage proposal to her, and the prospect of a wedding and a future together, however brief, seemed to engender a transformation of his outlook. Now properly motivated, he plunged into his doctoral work and began to grapple with issues of cosmic profundity.

AN INTRODUCTION TO BLACK HOLES

IN CHOOSING TO DO HIS PHD ON SOME OF THE GRANDEST THEMES IN COSMOLOGY — RELATIVITY, GRAVITY AND THE FABRIC OF SPACE AND TIME, AND THE LIFE CYCLE OF THE UNIVERSE — STEPHEN HAWKING REVEALED HIS MASSIVE INTELLECTUAL AMBITION.

Above: Roger Penrose, pictured in 1980. Penrose was a mathematician who had been switched on to cosmology by Dennis Sciama.

Opposite: A Penrose diagram: a way of representing the entire spacetime of the universe in two dimensions.

It was matched by intellectual fearlessness, illustrated by the tale of how he butted heads with the kingpin of British cosmology despite being just a sophomore postgraduate (*see* box).

Hawking's supervisor Dennis Sciama helped to engineer one of the most consequential meetings of Hawking's career, getting Stephen together with a brilliant mathematician, Roger Penrose. Penrose had much in common with Hawking: his father was also a professor in the life sciences, and he too had resisted pressure to follow in his father's footsteps. In the 1950s, Penrose had begun to forge new paths in cosmology by applying powerful mathematical methods to answer questions about relativity. When Hawking met him, he was using the branch of mathematics known as topology to achieve stunning proofs about black holes.

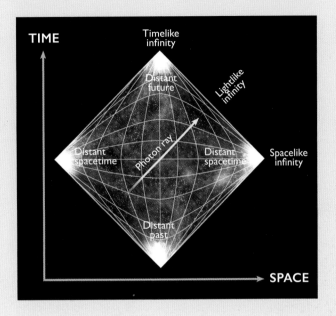

BUTTING HEADS WITH HOYLE

Hawking had struck up a friendship with Jayant Narlikar, the disciple of Fred Hoyle whose summer-school course had helped to cement his interest in cosmology. At Cambridge their offices were close to one another, and Narlikar generously allowed Hawking advance sight of a paper that he and Hoyle were co-authoring. The paper attempted to modify the theory of general relativity, to help reconcile Hoyle's steady state theory (see pages 22–3) with the steadily accumulating evidence against it. Hawking worked through the mathematics for himself and concluded that they did not add up. In June 1964, prior to the publication of the paper, Hoyle gave a presentation about it to the Royal Society in London. Hawking was in the audience, and when Hoyle had finished speaking he unsteadily got to his feet and pointed out that one of Hoyle's calculations was in error. Hoyle, who had no clue that this mere doctoral student he barely recognized had been afforded prior sight of the paper, was astonished, even more so by Hawking's reply when challenged as to how he could possibly know the calculation to be wrong: "I worked it out." To Hoyle and the assembled audience it seemed as though Hawking was claiming to have worked the fiendishly complex mathematics through in his head during the talk. It was a daring and potentially rash stroke in the still-deferential world of 1960s British academia. Hawking had begun to build his reputation as an intellectual heavyweight.

Prior to Penrose's work, black holes or singularities were widely considered to be an exotic and hypothetical outcome of certain ways of working through Einstein's field equations (the mathematical descriptions of general relativity). It was believed that only in certain very narrow and purely hypothetical circumstances could a star collapse into a singularity. By extending these circumstances Penrose offered mathematical proof that black holes might not only be more than a hypothetical curiosity, but might in fact be inevitable and even common in the real universe.

With topology, Hawking had stumbled upon a particularly appropriate mathematical approach that especially suited the mental visualization methods he was forced to adopt when his disability limited his ability to work through written equations (see pages 44–5). He would use Penrose's work as a jumping-off point for demonstrating that the concept of a singularity might hold the key to much more than black holes – perhaps to the origins of the universe itself.

WHAT IS A BLACK HOLE?

A black hole is the name given to a place where matter and energy have become concentrated into a small enough space for its gravitational attraction to become overwhelming, crushing it into a tighter and tighter ball. This in turn increases its gravity still further, shrinking it to a point of zero radius and thus infinite density: a singularity. Around the singularity is a region where the force of gravity is so strong that the velocity needed to escape it is greater than the speed of light. The boundary of this region is called the event horizon. Anything, including light, that crosses the event horizon cannot escape and is effectively cut off from the rest of the universe. Because no light can escape, reflect off or pass through the event horizon, it is utterly black – hence "black hole".

Below: An artist's depiction of a black hole, which has gathered around it a disc of material from a star it has torn apart.

UNDERSTANDING SPACETIME

Einstein showed with his theory of relativity that space and time are not independent of one another, and that in fact time is a fourth dimension alongside the three spatial dimensions that we normally recognize (height, breadth and depth). It has therefore become the convention to use "space-and-time", or simply "spacetime", as a more accurate way of describing the fabric of the universe. General relativity showed that matter and energy (which are actually two sides of the same coin) deform or warp the fabric of spacetime, changing its geometry, and that this geometry, which we know as gravity, in turn tells matter and energy how to move. American physicist John Wheeler, who coined the term black hole, summarized this

as: "Spacetime tells matter how to move; matter tells spacetime how to curve." The most popular analogy for understanding spacetime is a rubber sheet. The sheet represents four-dimensional spacetime; objects placed on the sheet represent matter/energy. So, a bowling ball placed on the sheet causes it to bow and curve, just as a mass of matter/energy causes spacetime to curve. The bigger the bowling ball, the deeper the resulting depression in the rubber sheet; similarly a massive object like a star causes greater curvature of spacetime than a smaller one like a planet. This kind of "depression" in spacetime is sometimes called a gravity well. A singularity is where the gravity well becomes infinitely deep – in other words a hole in the fabric of spacetime.

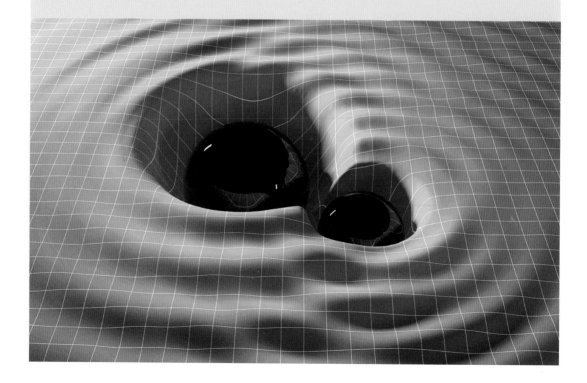

THE EVENT HORIZON

The event horizon of a black hole is not a thing – it's a place. The best-known analogy is with a boat on a river leading to a waterfall. As the boat draws nearer to the waterfall the current increases in strength, until at some point the current is stronger than the maximum power of the engine of the boat. Once it has crossed this threshold – or point of no return – nothing the captain of the boat can do will prevent the boat from going over the waterfall, but the captain himself will notice nothing different about the stretch of water before or after the point of no return, because there is no difference. Similarly, someone crossing the event horizon doesn't notice that they have crossed any boundary, and for them there is no difference in spacetime on either side of the event horizon. But once the threshold is crossed, the person can never escape the gravity of the black hole and will, inevitably, be drawn into the singularity.

Opposite: A graphic depiction of two black holes – represented as bottomless gravity wells in the fabric of spacetime – colliding and generating gravitational waves.

Above: An analogy for the event horizon around a black hole, where the black hole is the waterfall and the event horizon is the point – or rather border – of no return in the river.

A BRIEF HISTORY OF THE EXPANDING UNIVERSE

IN HIS PHD RESEARCH, STEPHEN HAWKING WOULD EXPLORE THE COSMOLOGY OF AN EXPANDING UNIVERSE AND ITS POSSIBLE ORIGINS. BUT HOW DO WE KNOW THE UNIVERSE IS EXPANDING AND WHAT DID COSMOLOGISTS WHO PRECEDED HAWKING THINK ABOUT ITS ORIGINS?

Above: A fanciful imagining of Hipparchus stargazing over the roofs of ancient Alexandria. In reality, he would not have had access to a telescope.

HERSCHEL AND GALACTIC LIFE CYCLES

The eighteenth- and nineteenth-century Anglo-German astronomer William Herschel built the world's most powerful telescope and used it to peer deeper into space than ever before. He spotted astronomical objects that he described as island universes (other galaxies), and was one of the first to speculate that galaxies might have life cycles. Of our own Milky Way, Herschel wrote that it "cannot last forever" and "its past duration cannot be admitted to be infinite".

Left: A reflecting telescope that Herschel himself designed and helped build. Using a device such as this, he discovered Uranus.

In ancient and medieval cosmology it was thought that the Earth is at the centre of the universe, surrounded by concentric spheres that contain the heavenly bodies (the Moon, Sun and planets), themselves surrounded by and contained within an outer shell on which the stars are set. Although not all the dimensions of this system were evident, the ancient Greeks used geometry to approximate the distances between the Earth and the Moon and Sun. In the second century BCE, for instance, Hipparchus calculated the distance between the Earth and Moon to be 59 times the radius of the Earth – a figure that is extremely close to modern estimates. Estimates of the Earth–Sun distance were less accurate, but it was apparent to ancient and medieval astronomers that the Solar System must be millions of kilometres in diameter.

Galileo was one of the first scientists to realize that the universe might extend far beyond the confines of the Solar System.

He realized that the stars, rather than being discs set in a shell or sphere at a fixed distance from the Earth, might be point sources of light set at varying and incomprehensibly vast distances. Armed with telescopes and freed from the dogma that the Earth is the centre of the universe, Enlightenment astronomers began to develop concepts of deep space.

Around a century later Einstein developed a new description of the universe, showing that space and time are relative and have curved geometry (*see* page 32), and that gravity is an expression of this curvature. The mathematical description of this theory is contained in Einstein's field equations, but the solutions to these appeared to show that the universe, rather than being static, is expanding. At the time Einstein assumed this to be a mistake, and to balance the effects of gravity he introduced into his calculations a factor known as the cosmological constant, which would produce a static solution.

$$G_{\mu\nu} + \Lambda g_{\mu\nu} = \frac{8\pi G}{c^4} T_{\mu\nu}$$

Above: Einstein's field equations, condensed into a single tensor equation, which describes how gravity results from the curvature of spacetime.

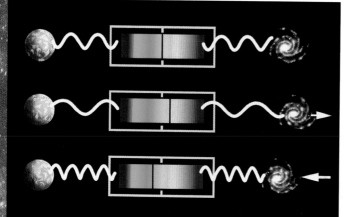

Above: The wavelength and thus colour of light travelling between galaxies depends on their relative motion: light from a galaxy moving away from us is shifted towards the red end of the spectrum, as in the middle picture.

REDSHIFT

American astronomer Edwin Hubble deduced the velocities of distant galaxies by measuring their redshifts. Redshift is the phenomenon by which the wavelength of light coming from a receding object is shifted towards the red end of the light spectrum. This is analogous to the Doppler effect, heard when an ambulance siren drops in pitch as it passes you and begins to move away. In this case, as the source of the sound waves moves away from you, the sound waves become stretched out. A longer wavelength means lower pitch. The same thing happens with light waves. As their source recedes so the light waves are stretched out, relative to a stationary observer, and so the wavelength of the light is redshifted. By measuring the degree to which light from distant galaxies is redshifted, Hubble was able to work out the speed at which they are moving away from us.

If Einstein was not ready to grapple with the implications of his theory, others were. In 1922, Russian physicist Alexander Friedmann was the first to demonstrate that Einstein's field equations had several valid solutions, including some in which the universe is expanding, and others in which it is contracting. Five years later, Belgian priest and physicist Georges Lemaître also worked through the field equations, demonstrating solutions that matched the observations of Edwin Hubble (*see* box), and concluding that the universe is expanding and that space and time had a starting point. In 1931 Lemaître published an English-language version of his theory, describing the starting point of the universe as the "hypothesis of the primeval atom" or the "Cosmic Egg". Lemaître talked about the Cosmic Egg "exploding at the moment of the creation", and described the beginning of time and space as a "now without yesterday".

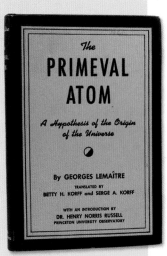

Right: The English-language version of Lemaître's book *The Primeval Atom*, one of his names for the cosmic singularity.

THE HUBBLE CONSTANT

Edwin Hubble was the pre-eminent astronomer of the twentieth century. In 1925 he cleverly used a class of stars of known brightness to prove that some were outside our own galaxies. This was one of the first definitive proofs that the universe extended beyond the Milky Way and that there were other galaxies in it. Four years later, Hubble presented the results of a decade-long survey of redshift measurements of distant galaxies, showing that all of them were moving away from us, and that the further away they were, the faster they were travelling. This correlation, known as Hubble's law, had universe-shaking implications. It suggested that the universe must be expanding, in turn suggesting that it had had a beginning, confirming the theories of Friedmann and Lemaître.

Left: Edwin Hubble looking through the eyepiece of the 254-cm (100-inch) telescope, at the Mount Wilson Observatory in Los Angeles, 1937.

A SINGULARITY IS INEVITABLE: HAWKING'S PHD

STEPHEN HAWKING TOOK PENROSE'S WORK ON BLACK HOLES AND USED IT TO MAKE A STUNNING BREAKTHROUGH IN THE MATHEMATICAL TREATMENT OF RELATIVITY AND THE BIRTH OF THE UNIVERSE.

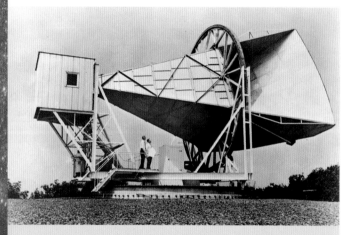

THE AFTERGLOW OF CREATION

When Hawking was working on his thesis there was no direct evidence for the Big Bang, but in the same year that he submitted it, radio astronomers Arno Penzias and Robert Wilson in the United States picked up a peculiar hiss with their telescope. Through patient and diligent detective work they showed that this background hiss is not of earthly origin, and that it represents a very faint and cool trace of heat spread across the entire universe. Although today this heat is just a few degrees above Absolute Zero, it proves that billions of years ago, when the universe was much smaller, it must have had a temperature of millions of degrees. What Penzias and Wilson had picked up is the residual heat of the Big Bang, sometimes described as the "afterglow of Creation". In 1978 the pair were awarded the Nobel Prize in physics for their discovery.

This incredible achievement formed the fourth and final chapter of his PhD thesis, which sets out its stall in the first line of its abstract: "Some implications and consequences of the expansion of the universe are examined."

As detailed on pages 34–7, there is ample evidence that the universe has expanded and continues to do so. When Hawking was writing his PhD thesis, there was also already a shaky theoretical basis for believing this. The simplest solution to Einstein's field equations uses what is known as the Robertson-Walker metric, in which metric means a way of measuring spacetime. With this metric, space can change over time, expanding as time runs forwards. Running the equations with the clock in reverse describes a universe that contracts, until it ultimately reaches a point – the cosmological singularity that Lemaître called the Cosmic Egg (*see* page 37). However, steady state exponents such as Hoyle were able to dismiss this solution to the field equations as an error; a flaw in the theory of general relativity.

Hawking made use of the mathematical methods pioneered by Penrose, who had shown what happens when a massive star collapses and its mass contracts into a point. By reversing

Left: The Holmdel Horn Antenna at Bell Telephone Laboratories, which Penzias and Wilson used to detect the afterglow of the Big Bang.

time's arrow in his calculations, Hawking described this kind of contraction in reverse: an explosion starting from a singularity. Hawking was able mathematically to prove that such a process is not only theoretically viable but "that a singularity is inevitable, provided that certain very general conditions are satisfied". In his doctoral thesis Hawking had demolished one of the primary arguments for the steady state model and against the Big Bang model, showing the Big Bang is a necessary consequence of general relativity and that the origin of the universe could very well have arisen from a cosmological singularity.

WHAT ELSE WAS IN HAWKING'S PHD?

The proof of the cosmological singularity comes in the final chapter of Hawking's PhD. In the first chapter he explored flaws in the Hoyle-Narlikar theory of gravitation, relating to his infamous criticism of Hoyle at the Royal Society in 1964 (see page 29). In Chapter 2, he examined how galaxies might have been formed in an expanding universe and looked at what he called gravitational radiation, better known today as gravity waves. Chapter 3 explores the mathematics of gravity waves in more detail.

Above: A picture of the whole sky, showing the cosmic background microwave radiation – the very faint glow of heat left over from the Big Bang, which preserves a record of the universe as it looked about 13.7 billion years ago.

SOMETHING TO LIVE FOR: MARRIED WITH CHILDREN

SOME OBSTACLES STOOD IN THE WAY OF STEPHEN AND JANE'S MARRIAGE. FOR ONE THING, JANE'S COLLEGE, WESTFIELD, DID NOT PERMIT ITS STUDENTS TO GET MARRIED.

Compounding this problem was the fact that Jane's father had made it a precondition that she should finish her degree. In the light of Stephen's condition and uncertain life expectancy, the college relented, although Jane would have to continue to study in London, without sharing a room, even after they were married. On Stephen's side, the main problem was that, in order to consider marriage, he needed a job. In 1965, with his PhD well underway, Stephen applied for and won a research fellowship at Gonville and Caius college in Cambridge. He also won money for an essay he submitted to the Gravity Prize Competition. The way was now clear for the young couple to wed.

At the wedding on 14 July 1965, a civil-marriage ceremony conducted by the registrar at the Shire Hall in Cambridge, Jane faced her new parents-in-law with mixed feelings. Stephen's mother had not given her the friendliest of welcomes, telling her, "We don't like you because you don't fit into our family." Meanwhile his father, with what Jane described as characteristic Yorkshire forthrightness, had advised her to get on with having children, since Stephen's life would be short and his ability to "fulfil his marital duties" more short-lived still. Jane chose not to learn the full details of Stephen's ghastly prognosis, explaining, "I did not see much point in having whatever natural optimism I could muster destroyed by a litany of doom-laden prophecies.... I loved Stephen so much that nothing could deter me from wanting to marry him." For Stephen's part, he later acknowledged that the marriage gave him "something to live for".

Back home in Cambridge the young couple struggled to find suitable housing, and at first Jane continued to spend all week in London working on her degree. However, they soon started a family, with their children Robert and Lucy arriving in 1967 and 1970 respectively. A third child, Timothy, was born in 1979.

Opposite: Jane and Stephen on their wedding day, flanked by their parents. In fact they had two wedding days; a civil ceremony on 14 July, at a registry office, followed the next day by a religious ceremony at Trinity Hall chapel.

Above right: Jane and Stephen with their children Robert and Lucy, a few years before the arrival of their third child, Timothy.

NEWLYWEDS ABROAD

On 15 July 1965 there was a religious marriage ceremony in the chapel at Trinity Hall, after which Jane and Stephen climbed into the red Mini they had just bought and set off for a short honeymoon in Suffolk (a rural part of eastern England). Just one week later they were boarding a plane to carry them across the Atlantic so that Stephen could attend a summer school on general relativity at Cornell University in upper New York state. Although he was excited to be mixing with internationally renowned physicists, the trip put immediate strains on the marriage. This was because Jane was forced to cope with unsuitable lodgings, while trying to pursue her own research. Of course, she also wished to be a supportive housewife, but this was a task made more difficult by a dramatic illustration of the severity of Stephen's condition that took place soon after the wedding. Stephen suffered a severe choking fit, from which Jane rescued him by means of a hearty thump on the back; this scary incident graphically brought home to her the health obstacles that they would face together in the future.

CHAPTER 3

THE BLACK HOLE REVOLUTION

WHAT NOBODY ELSE CAN DO: SCIENCE WITH A DISABILITY

ALTHOUGH STEPHEN'S ALS (OR MOTOR NEURONE DISEASE) HAD NOT PROVED TO BE AS DANGEROUS AS INITIALLY FEARED, HIS PHYSICAL DEGENERATION CONTINUED.

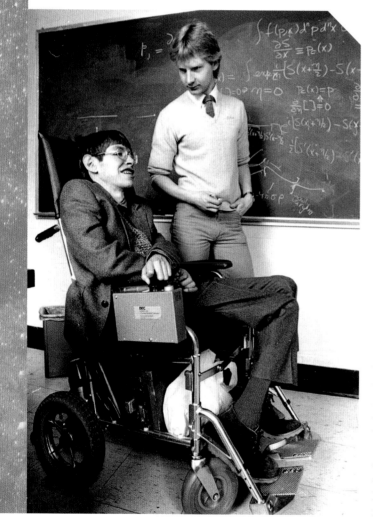

Around the house and in the office his iron will and immense stubbornness manifested themselves in a refusal to let others help him when it was remotely possible for him to do something for himself. Every evening, for instance, he would painfully and slowly pull himself up the stairs to go to bed. For Jane and others who had to help him, this refusal to make concessions to his illness made life much harder.

By 1970 Stephen was forced, reluctantly, into a wheelchair. His speech was slurred at best, making lecturing practically impossible, and he could not write on paper or on a blackboard, making ordinary mathematics extremely difficult to practise. However, by a happy accident Stephen had chosen one of the few professions in which ALS need not be a serious barrier to accomplishment, since his branch of cosmology is an almost entirely intellectual rather than physical pursuit. His disability forced him to develop a new way of working, with a unique ability to visualize four-dimensional spacetime, which no one else in physics had, and which well-informed observers such as Kip Thorne and Isobel Hawking say he would never have developed without his ALS.

Thorne, for instance, noted that Stephen developed "geometrical arguments that he could do pictorially in his head", equipping him with "a very powerful set of tools that nobody else really had", which meant that

Opposite: Hawking with a graduate student;
he came to rely on graduate students to interpret
his speech and write out calculations for him.

Below: The building at Gonville and Caius college
that housed Hawking's office.

there were "certain kinds of problems" that only Hawking could solve. Hawking himself described equations as the "boring part of mathematics", insisting that he preferred to "see things in terms of geometry". His mother relates that Stephen himself claims "that he wouldn't have got where he was if he hadn't been ill".

A consequence of his disability was to free Hawking from distractions and duties that might have sapped his time and focus. Since he could not be expected to help with any domestic or professional duties, he had the luxury of thinking purely about physics. According to Isobel Hawking, "he has concentrated on [physics] in a way I don't think he would have done otherwise.... I can't say anyone's lucky to have an illness like that, but it's less bad luck for him than it would be for some people, because he can live so much in his head."

NIGHT THOUGHTS

A good illustration of the way that Hawking was both forced and freed to think hard about his subject was the way in which his 1970 discovery of the second law of black hole dynamics was made. The initial inspiration for this occurred to him as he was engaged in the painstaking and lengthy process of getting ready for bed. Rather than going to his desk to start writing, as an able-bodied person would have done, Hawking lay awake all night thinking it through, waiting until he could phone Penrose with his findings in the morning.

Left: An artist's interpretation of two black holes merging. A process, Hawking realized, that must increase the total entropy.

BLACK HOLE DYNAMICS

IN CAMBRIDGE, HAWKING CONTINUED TO WORK ON THE ASTROPHYSICS OF BLACK HOLES.

In 1970, again working with Roger Penrose, he devised what he called the second law of black hole dynamics, in a deliberate reference to the second law of thermodynamics. The latter is one of the most fundamental laws of nature. It sets what physicists call "the arrow of time", which is to say the direction of time, from past to future, and it states that entropy always increases, and can never decrease. Entropy is a difficult concept to grasp, and is variously defined as disorder, waste heat or loss of information. Hawking's law of black hole dynamics said that black holes, like entropy, cannot decrease. Or to put it another way, a black hole can never get smaller.

Hawking proved this theory by considering the fate of rays of light near the event horizon, and realizing that shrinkage of the event horizon would mean these rays converging, colliding and scattering into the black hole, so that in fact the event horizon would not shrink. In other words, the event horizon can only

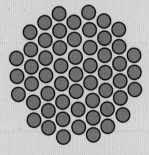

Highly ordered

Increase in entropy

More disordered

Above: An example of increasing entropy or disorder. The mosaic sphere on the left is highly ordered, but once smashed into hundreds of fragments it becomes highly disordered.

shrink if the black hole loses mass, and since nothing can ever get out of the black hole, this should be impossible. All that can occur is for matter to cross the event horizon and fall into the black hole, in which case it grows and the event horizon becomes wider.

Hawking knew his law of black hole dynamics to be closely analogous to the second law of thermodynamics. There appeared to be a direct parallel between entropy and the size of the event horizon. The second law of thermodynamics states that when two systems are combined, the entropy of the combined system must be as great or greater than the sum of the original systems. For instance when a hot cup of tea (low entropy) is mixed with a cold cup of tea (high entropy), the heat of the combined cup must be the same as or lower than the sum of the temperatures of the

THE RUMOUR MILL

Hawking's position at Cambridge, working in the Department of Applied Mathematics and Theoretical Physics (DAMTP), depended on one of the colleges employing him, but his fellowship at Gonville and Caius (G&C) was drawing to an end. To ensure that he was offered a new one, Hawking's friends conspired to pull off a neat ruse. Dennis Sciama and Hermann Bondi colluded to create a rumour that Hawking was about to be poached by King's College in London, prompting G&C quickly to offer him a new, specially created fellowship that took account of the fact that Stephen could no longer give lectures. His value as a thinker was being recognized.

SECOND LAW OF THERMODYNAMICS

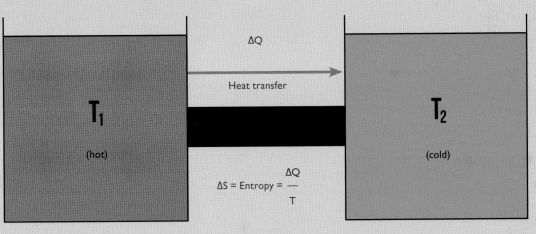

ΔQ

Heat transfer

T_1

(hot)

T_2

(cold)

$$\Delta S = \text{Entropy} = \frac{\Delta Q}{T}$$

Above: A system (two containers joined together) displaying the second law of thermodynamics in action. As heat energy dissipates from the hot container to the cold one, the total entropy of the system increases.

WHEELER'S ETERNAL CRIME

American physicist and "father" of the black hole John Wheeler told an amusing anecdote about how he inspired Israeli-American physicist Jacob Bekenstein to challenge Stephen Hawking. He related musing to Bekenstein that, when he put a hot and cold teacup together, allowing heat to flow from one to the other and in the process increasing the amount of disorder (that is, entropy) in the universe, he had "committed a crime that will echo down the corridors of eternity". However, he observed, if he were able to drop both cups of tea into a convenient black hole, he would have "concealed the evidence" of his crime. This is a colourful way of illustrating the argument that a black hole is cut off from the rest of spacetime and thus not really part of our universe, so that entropy that "falls" into a black hole is "lost" to the universe, breaking the second law of thermodynamics. If Hawking was right, therefore, and black holes cannot really have entropy, he had created the paradox of decreasing the entropy of the universe. As Wheeler recalled, Bekenstein "looked troubled" and came back later to inform him that in fact the black hole did record the evidence of his crime, and that the black hole *had* increased its entropy. This set him on a direct collision course with Hawking.

Left: John Wheeler, who pioneered the theory of black holes and came up with their evocative name.

individual cups (and therefore its entropy the same or greater). Similarly, Hawking's law of black hole dynamics showed that bashing together two black holes cannot result in either of them being broken apart or becoming smaller; instead they will combine to create one as large or larger than before.

But Hawking fervently contended that the analogy was just that – black hole dynamics and thermodynamics are analogous but not identical – and he had good reasons for arguing this. Recall that one definition of entropy is waste heat (aka lost or unavailable energy); this means that a body with entropy must have some amount of heat: it must have a temperature. However, in order to have a temperature a body must be radiating heat energy, and a black hole cannot, by definition, radiate heat energy because no radiation can escape from inside the event horizon. Admitting black hole dynamics

Above: Hawking's mind ranged across the cosmos, exploring the most extreme conditions in the universe.

PRIMORDIAL BLACK HOLES

In the present-day universe, black holes form when stars run out of fuel, so that the outward pressure of radiation generated by their fusion no longer counterbalances the inward pull of gravity, and they collapse. Provided that the star is of sufficient mass (around three times the mass of our sun), the force of gravity generated is great enough to exceed the escape velocity of light, and a black hole results. This means that all "modern" black holes must be at least three solar masses in mass. In 1971 Stephen Hawking showed that in the extreme conditions of the early universe, when it was less than a second old, much smaller masses could collapse into black holes. The hotter the universe at the time, the less mass was needed to create a black hole, so that very soon after the Big Bang, very tiny black holes could have been created. These primordial black holes would thus have very different properties to "modern" ones, and they would also be extremely ancient (around 13.5 billion years old). This would prove to be highly significant in the light of Hawking's next discovery.

Below: Artist's conception of the Big Bang, with all of space, time, matter and energy exploding forth from a single point.

and thermodynamics to be more than merely analogous would create a paradox. This was the position set forth by Hawking when he announced his new law of black hole dynamics at the Texas Symposium of Relativistic Astrophysics in December 1970. However, some of those attending were concerned that by ruling out one paradox, Hawking was allowing another: what John Wheeler described as covering up "a crime that will echo down the corridors of eternity" *(see* box).

Jacob Bekenstein's response to Hawking's presentation in Texas was therefore to insist that thermo- and black hole dynamics are more than just analogous: they are the same thing. Black holes do indeed have entropy, and the area of the event horizon is the entropy of the black hole. The Englishman was annoyed, fiercely resisting Bekenstein's interpretation with its attendant paradox, and continuing to do so even while working with James Bardeen and Brandon Carter to formulate a whole set of laws of black hole dynamics. They came up with four laws that directly paralleled the four laws of thermodynamics, but with the terms "horizon area" and "horizon surface gravity" substituted for entropy and temperature, respectively. Yet they insisted that the two sets of laws were merely analogous and not identical; a black hole could not, by definition, have a temperature. Bekenstein retained his doubts.

Perhaps prompted by this challenge, from 1973 Hawking would begin to explore an entirely new realm of physics in his quest to gain a fuller understanding of black holes. In the process he would attempt the first synthesis of the two cornerstones of modern physics, relativity and quantum mechanics: a theory of quantum gravity.

Above: Jacob Bekenstein in later life, in his office at the Hebrew University in Jerusalem in 2009.

RELATIVITY VS QUANTUM MECHANICS

Hawking was about to embark on perhaps the toughest quest in modern physics, one that defeated Einstein and many others, as he sought to reconcile relativity and quantum mechanics, two apparently contradictory models of reality. Relativity is the theory that describes how the universe works on a cosmic scale, and shows that space and time cannot be separated and are part of a single, seamless fabric. Quantum mechanics is the theory that describes how the universe works at a subatomic scale, and it shows that space and time cannot be seamless and must be separated into discrete chunks. In a sense, relativity is analogue and quantum mechanics is digital.

ABSOLUTE RUBBISH: HAWKING RADIATION AND EXPLODING BLACK HOLES

IN AUGUST 1973, KIP THORNE ACCOMPANIED THE HAWKINGS ON A TRIP TO POLAND, IN ORDER TO INTRODUCE STEPHEN TO A PAIR OF SOVIET PHYSICISTS.

Yakov Zel'dovich and Alexander Starobinsky had recently shown that the uncertainty principle (*see* box) suggests that the rotational energy of a spinning black hole can cause particles to be created and emitted. Hawking wanted to explore this further, and replicate with his own mathematical treatment the Russians' result. To his surprise, he discovered that the calculations seemed to show "that even non-rotating black holes should create and emit particles at a steady state". Annoyingly, this finding appeared to verify Bekenstein's claims

about black holes having entropy, and thus being hot (or at least, slightly warm) objects.

Hawking brooded over his result for months. It looked as though Bekenstein might have been right after all, and Hawking was not pleased. How could it be that a black hole really did give off particles? The answer, he would show, lies with concepts such as the uncertainty principle, virtual particle pairs and the weird world of the event horizon.

With an elegant treatment of quantum mechanics in the very special space-time

UNCERTAINTY PRINCIPLE

Formulated by Werner Heisenberg in 1927, the uncertainty or indeterminacy principle states that it is impossible to be certain of both the position and the movement of a particle at the same time. If you measure its position, you cannot know how it is moving, and if you measure how it is moving, you cannot know its position. The more you try to pin down (or determine) one of these complementary aspects of the nature of the particle, the less certain (or more indeterminate) the other aspect becomes. The principle applies to energy fields as well as particles, so that it is impossible to know both the strength of a field and its rate of change. This has profound consequences for the notion of a vacuum or of space itself.

Left: German physicist Werner Heisenberg at work, possibly on his uncertainty principle, although it's hard to be sure.

Above: Hawking's theoretical black hole radiation may not be proven until it is possible to send probes close to the event horizon of a black hole, as in this concept image.

environment of the black hole, Hawking was able to show that the intense gravity of the event horizon "promotes" virtual particle pairs into real particle-antiparticle pairs. While the virtual particle pairs have a combined and inseparable net energy of zero (described as the pair being $+-E$), at the event horizon they are promoted into an outgoing particle with energy $+E$ paired with an in-falling antiparticle with energy $-E$.

Another way of looking at this is to say that the event horizon of a black hole acts like a razor blade, cutting the ties that bind virtual particle pairs to their fate of instantaneous mutual annihilation, so that

one of the pair crosses the event horizon and falls into the black hole. This severs its link with the other virtual particle, which is now free to fly off into space as a real particle. These decoupled virtual particles are the source of the radiation that Hawking had discovered must be emitted by black holes, which would come to be known as "Hawking radiation". Hawking radiation resolves both the paradoxes of black hole dynamics: the apparently lost entropy that troubled Bekenstein, and the emission of radiation from a body that cannot radiate. Hawking radiation means that black holes can have entropy. It also had startling consequences for the fate of black holes.

NO SUCH THING AS SPACE

The uncertainty principle states that there is no such thing as empty space. A true vacuum containing no matter or energy fields would violate the principle, since the strength of a field and its rate of change would both be zero; both properties would be determinate, and we know from the indeterminacy principle that this is impossible. In fact, at the quantum scale, space is not empty; because the properties of fields are indeterminate or fuzzy, a vacuum is actually filled with constantly fluctuating fields and the particles that mediate these fields (particles such as photons and gravitons). Pairs of complementary and opposing particles are constantly appearing for the briefest of moments and then almost instantaneously annihilating one another. Each pair of virtual particles consists of a particle and its anti-particle, one of which has positive energy and the other negative. In this way the energy balance of the universe is conserved since the positives and negatives balance out. These particles are known as virtual particles, because they pop into and out of existence so quickly that they make little or no impact on the macroscopic world.

Below: A conceptual artwork of a virtual particle pair springing into existence, in the process known as vacuum or quantum fluctuation.

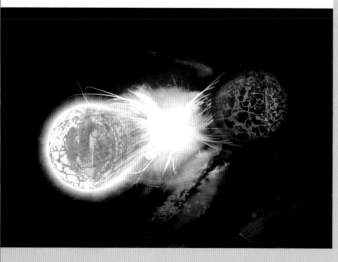

The fact that black holes can give off Hawking radiation means that they can lose mass, shrink and eventually evaporate. Again this seems paradoxical – how can a black hole lose mass, given that nothing can escape it? The explanation for this is that when a pair of virtual particles is separated by the event horizon, the one that falls into the black hole is the negative energy particle. Swallowing a negative energy particle means that energy (and therefore mass which, as Einstein showed, is equivalent to energy) is subtracted from the black hole.

As the black hole loses mass through "energy subtraction", its event horizon will shrink. The rate of shrinkage depends on how much radiation is coming off the black hole, and this in turn depends on its size. The bigger the black hole, the greater its entropy and the lower its temperature, and thus the smaller the amount of radiation emitted and the more slowly it will shrink. Only with small black holes will there be a measurable amount of radiation, and the only small black holes we know of are the primordial black holes that Hawking had predicted would have formed in the early universe (*see* box, page 50). In fact, he calculated, these tiny black holes would glow white hot with radiation, emitting so many particles that they would rapidly shrink, getting hotter and smaller and smaller and hotter: a vicious circle that means that subatomic-size black holes will explode with colossal energy.

These were stunning findings that appeared to contradict everything that was known about black holes. Hawking was extremely concerned that he must have made a mistake, and was reluctant to share his results, discussing them with only a few friends. He was not impressed to receive a phone call from an excited Roger Penrose, who had got wind of his findings, just as he was sitting down for dinner. As he later lamented, his dinner grew cold: "It was a great pity, because it was goose, which I'm very fond of."

Penrose was not the only one who was worked up. Hawking's friend, the astronomer Martin Rees, who had been one of the select few that Hawking briefed on his work, met with Dennis Sciama in Cambridge early in 1974. Sciama recalls that Rees was "shaking with excitement", exclaiming "Everything is different, everything has changed!".

The response of some others in the cosmology community was less enthusiastic. Hawking presented his bombshell findings at Sciama's 1974 Second Quantum Gravity Conference, in a paper entitled "Black Hole Explosions?". The question mark was added at the last moment as Hawking looked to get his defence in early. His presentation was received in bemused silence, until the moderator John G. Taylor got up and said, according to one account, "Sorry Stephen, but this is absolute rubbish."

Above: Sir Martin Rees, a friend of Hawking's since they had both worked with Sciama at Cambridge; he later became Astronomer Royal.

REACTIONS TO HAWKING'S NEW THEORY

In spite of the poor reaction to his latest theories at the 1974 Second Quantum Gravity Conference, the power and clarity of Hawking's vision would not be denied. He published his paper in the prestigious scientific journal *Nature* in March 1974, and the scientific community had no choice but to accept its findings. When Kip Thorne met with Zel'dovich and Starobinsky in Moscow, they greeted him with hands aloft in surrender. Sciama called the paper "one of the most beautiful in the history of physics", while John Wheeler described it as like "rolling candy on the tongue".

COMING TO AMERICA

BY 1974, STEPHEN HAWKING COULD NO LONGER MAKE THE LABORIOUS NIGHTLY ASCENT OF THE STAIRS TO HIS BEDROOM. THE HAWKINGS MOVED TO A GROUND-FLOOR FLAT, BUT THE DEMANDS ON JANE'S TIME AND PSYCHE WERE BECOMING IMPOSSIBLE.

One welcome redress came when they adopted the practice of having a graduate student to also help Stephen in the role of part-time carer. Bernard Carr was the first to fill this post.

In the spring of 1974 Hawking was given the honour of being inducted as a Fellow of the Royal Society – one of the oldest scientific institutions in the world – of which Isaac Newton had been an early president. Around the same time, Srephen and his wife Jane accepted an offer arranged by Kip Thorne, for Hawking to become a Sherman Fairchild Distinguished Scholar at the California Institute of Technology (aka Caltech) in Pasadena, where they would spend an academic year.

Jane, Stephen, their two children and their live-in graduate student crossed the Atlantic and arrived in California in August 1974. Waiting for them were a new car, a beautiful home and an electric wheelchair that delighted Hawking and would become one of his signature accessories. For Jane, California proved to be a social whirl, and she was introduced to a pastime that would become a passion, joining a chorus.

For Stephen, Caltech was a fertile breeding ground of ideas and collaborations. He attended lectures given by two of the world's leading particle physicists, Richard Feynman and Murray Gell-Mann. He also worked with American physicist Jim Hartle on developing his theory of Hawking radiation, and published a paper with an American graduate student, Don Page, on primordial black holes exploding as gamma ray bursts. All in all, Hawking's time as a visiting scholar at Caltech was deemed such a success that he would return nearly annually to spend a month there.

Left: The entrance to the Royal Society, in its Carlton House Terrace home in London, where it has been since 1967.

BY PAPAL INVITATION

In 1975 Hawking was invited to Rome to receive the Pius XI Medal, awarded by the Pontifical Academy of Sciences every two years to a young scientist under the age of 45, chosen for his or her exceptional promise. He met Pope Paul VI, and would go on to meet three other popes, including Pope Francis (papacy 2013–present). While in Rome, Hawking made a point of visiting the Vatican Library to view Galileo's recantation of his heliocentric heresy. Not long after this trip the Vatican formally apologized for its treatment of Galileo.

Top: The eccentric and brilliant American theoretical physicist Richard Feynman, pictured in 1954.

Above: Part of the California Institute of Technology, known as Caltech.

A SCIENTIFIC WAGER

THERE IS A LONG TRADITION IN SCIENCE, AND PARTICULARLY IN COSMOLOGY, OF MAKING BETS OR WAGERS ABOUT THEORIES AND WHETHER THEY WILL BE PROVED OR DISPROVED.

Stephen Hawking, with his impish sense of fun and flair for the dramatic, embraced this tradition wholeheartedly, becoming perhaps its greatest modern practitioner. In the course of his career he made many high-profile wagers, one of the earliest and best known of which was his 1974 wager with the American physicist Kip Thorne over the existence of a particular black hole.

In 1964, astronomers discovered a strange astronomical object in the constellation Cygnus. Labelled Cygnus X-1, this unknown body emits colossal blasts of X-rays into space. Cosmologists proposed that it must be a black hole that is sucking in huge quantities of gas and dust, and that the vast gravitational pull of the black hole accelerates this matter to near-light speed, causing it to become super-heated and give off intense radiation. Hawking, who had invested his entire career in the existence of black holes, badly wanted this prediction to be right, and he and Thorne, a colleague during his time at Caltech, amused themselves by cooking up a wager.

Written out by hand, presumably by Thorne, the wager specifies that:

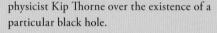

Left: Cygnus X-1, as captured by X-ray telescope Chandra.

Whereas Stephen Hawking has such a large investment in General Relativity and Black Holes and desires an insurance policy, and whereas Kip Thorne likes to live dangerously without an insurance policy. Therefore be it resolved that Stephen Hawking bets one year's subscription to "Penthouse" as against Kip Thorne's wager of a 4-year subscription to "Private Eye", that Cygnus X 1 does not contain a black hole...

At the bottom of the wager are two signatures, one of them extremely shaky.

Although Cygnus X-1 is unlikely ever to be definitively proven to be a black hole, in 1990 Hawking considered the evidence strong enough to conclude that he had lost the wager. On a visit to Los Angeles, according to Thorne himself, Hawking broke into his office and signed off on the bet by affixing an inky thumbprint to the bottom of the document. Hawking then duly fulfilled his half of the bargain by buying Thorne a subscription to the racy soft-porn magazine *Penthouse*, noting that the American's wife was not at all amused by his actions.

Below: Text of the 1974 wager between Kip Thorne and Stephen Hawking, with Hawking's shaky signature at bottom left.

CHAPTER 4

THE BLACK HOLE WAR

KEYS TO THE UNIVERSE: RISING PROFILE

IN THE SUMMER OF 1975 THE HAWKINGS RETURNED TO ENGLAND, TO CAMBRIDGE, A BASE FROM WHICH STEPHEN WOULD GO ON TO LAUNCH A CAMPAIGN OF GLOBAL CONQUEST.

Advancements in his career and academic prestige were matched by his growing public profile, as for the first but by no means the last time, trucks and cameras descended on the Hawking household to produce a documentary. The 1977 BBC documentary *The Key to the Universe: The Search for the Laws of Creation*, a survey of contemporary advances in particle physics and cosmology, would mark the initial steps in transforming Hawking into first a national and later a global figure. It brought into the public consciousness the extraordinary juxtaposition of a great mind, roaming the cosmos, with a frail and twisted body: "Although the gentle gravity of the planet Earth confines him to a wheelchair," intones the narrator, "in his mind he masters the overwhelming gravity of a black hole."

As Hawking's condition worsened, steps were

Above: The Department of Applied Mathematics and Theoretical Physics (DAMTP) on Silver Street in Cambridge.

Opposite above: Hawking in his office at Gonville & Caius with Judy Fella, 1981.

taken to improve his working arrangements. First he secured a readership to succeed his expired fellowship, enabling him to continue at the Department of Applied Mathematics and Theoretical Physics (DAMTP). Next it was arranged for him to have a full-time secretary, with the dynamic Judy Fella coming in to relieve Jane of some of her administrative burdens. Fella also significantly smoothed Stephen's path in academia, for instance helping to organize his constant visits to other educational institutions and scientific conferences. In the autumn of 1976, Don Page, with whom Stephen had worked so effectively at Caltech, came over to join the Hawking household as their resident graduate student. Page was a committed evangelical Christian – apparently an odd match for the atheistical Hawking – but Stephen seemed to enjoy gentle and good-natured theological sparring.

WEAPON OF CHOICE

When they first met, Jane had been shaken by Stephen's reckless driving style, which appeared to make no allowances for his disability. His electric wheelchair now gave him a second chance to live dangerously. He became notorious for his cavalier disregard for caution, for example gleefully plunging at full speed down steep San Francisco streets, or carelessly toppling off the edge of the stage while giving a presentation; on one such occasion he quipped that he had "fallen off the edge of the universe". Stephen was also not above using his wheelchair as a weapon, or at least as a means with which to express mild contempt. In one infamous example in 1977, invited to attend the induction of Prince Charles to the Royal Society in London, Stephen contrived to run over the Prince's toes. He was later said to have lamented never getting the chance to do the same to Margaret Thatcher, although he dismissed the accusation that he had ever used his wheelchair in this way as "a malicious rumour – I'll run over anyone who repeats it".

Right: Hawking was delighted with the autonomy and freedom afforded by his electric wheelchair.

LUCASIAN PROFESSOR

IN THE AUTUMN OF 1977 STEPHEN HAWKING WAS MADE A PROFESSOR, WHILE AN ASSORTMENT OF AWARDS AND ACCOLADES BEGAN TO MOUNT UP. IN 1976 HE WAS AWARDED THE ROYAL SOCIETY'S HUGHES MEDAL, GIVEN TO "AN OUTSTANDING RESEARCHER IN THE FIELD OF ENERGY", AND IN 1978 HE WON THE HIGHEST ACCOLADE IN AMERICAN PHYSICS, THE ALBERT EINSTEIN AWARD.

Hawking was given honorary doctorates by many academic institutions, including Oxford University. In 1979 he was made Lucasian Professor of Mathematics, perhaps the most prestigious chair in British – and arguably world – science and one with a prestigious history stretching back to the seventeenth century (*see* box). Hawking would hold this post for the next 30 years. It is customary for those awarded such distinctions at Cambridge to sign their names in a ledger. Hawking's shaky effort on acceptance of this great accolade would be the last time that he signed his name.

By this time Stephen Hawking had set his sights on accomplishing Einstein's dream – "a complete and consistent theory that would unify all the laws of physics". In 1980, his inaugural lecture as Lucasian Professor was ambitiously titled "Is the End in Sight for

MATHEMATICK PROFESSORS

Named after Henry Lucas – a distinguished clergyman who left Cambridge University his library and a grant of land to fund a chair in "mathematick" – and formally established by Charles II in 1664, the Lucasian Professorship has been held by some great names in science. The first holder, Isaac Barrow, did important work on calculus that would be taken up by his protégé and successor in the chair, Isaac Newton. Other distinguished holders of the post have included Charles Babbage, father of the computer, and Paul Dirac, Nobel prize-winning theoretical physicist.

Right: Isaac Barrow, first holder of the Lucasian Professorship of Mathematics; a mentor of Isaac Newton, he did important work on optics and calculus.

Theoretical Physics?". In it he set out his belief that a Theory of Everything is possible, and pointed, as the most promising candidate, to the theory known as N=8 supergravity (*see* "Supersymmetry", page 66).

Below: A conceptual artwork depicting elements of a Theory of Everything, including fundamental forces and particles, space and time, and Einstein's field equations, amongst others.

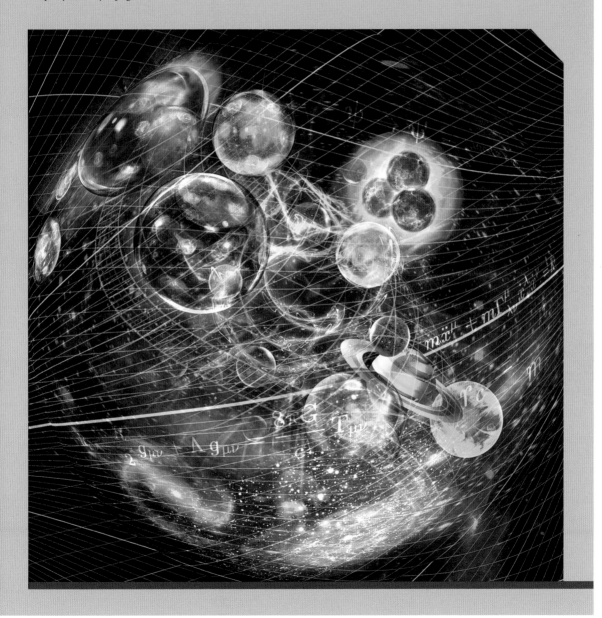

SUPERSYMMETRY

Hawking's 1980 candidate for Theory of Everything is a variant of the theory known as supersymmetry, itself an attempt to achieve a Grand Unified Theory (*see* box opposite) and reconcile flaws and gaps in the Standard Model of particle physics. To understand supersymmetry it is necessary to understand what symmetry means in terms of fundamental physics. Symmetry in physics refers to natural laws operating in predictable ways within certain constraints. For instance, a basic symmetry of nature is that the laws of physics are the same throughout space and time, so that a physics experiment will produce the same result wherever and whenever it is performed. The laws of physics are also symmetrical with respect to speed, a fact that led Einstein to realize his theory of special relativity: this symmetry showed him that space and time are relative. Space-time symmetries at the level of fundamental physics govern the

interactions and properties of fundamental particles, and make it possible to predict the existence of unobserved particles.

But what if there are more dimensions of space and time than those we can observe? Particle physics theorists realized that symmetries in these higher dimensions might explain a lot of the things in the Standard Model that currently do not make sense, and fill in some of the gaps. They call these higher-dimensional symmetries "supersymmetries". In the theory of supersymmetry, each particle in the Standard Model will have a supersymmetrical counterpart that helps to explain things in the Standard Model that are inconsistent or do not make sense. For example, the particle that mediates gravity, the graviton, would have a supersymmetrical counterpart called the gravitino. In Hawking's preferred "flavour" of supersymmetry, N=8 supergravity, there are actually eight types of gravitino.

If supersymmetry is correct, there are dozens more fundamental particles out there, so why haven't we discovered any? If the superparticles are genuinely symmetrical with known particles, they should have similar mass and thus similar properties, and we should know about them. Clearly this is not the case, and physicists think the reason is that some property of our universe is responsible for what they call "symmetry-breaking", although "symmetry-masking" might be a better term. This property, which is probably something to do with the Higgs field that gives mass to

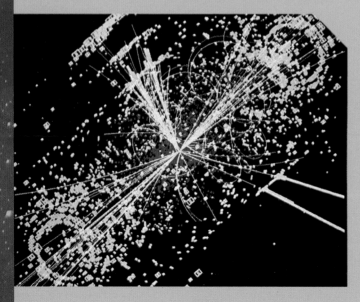

Left: An example of the kind of data gathered by the Large Hadron Collider, showing particles' paths as protons are collided, creating smaller particles including a Higgs boson.

particles, has the effect of masking the existence of superparticles by increasing their mass to levels beyond the capacity of present-day technology to create and/ or detect. However, it is possible that the Large Hadron Collider will be able to generate high enough energy levels to detect some forms of superparticle, just as it managed to detect the extremely massive Higgs boson (*see* page 111).

Left: A conceptual illustration of aspects of supersymmetry, in which known particles may have counterparts in higher dimensions.

ToE VS GUT

The holy grail of theoretical physics is to combine different theories or models of the universe into unified syntheses. However, the terminology of this ambition can become confusing, especially since there is a distinction between the Grand Unified Theory (GUT) and the Theory of Everything (ToE). The GUT is the hoped-for unification of the different aspects of the Standard Model of particle physics (the account of the fundamental particles and forces of nature), with its three fundamental forces: electromagnetism and the strong and weak nuclear forces. In fact, physicists have already achieved a unification of the theory of electromagnetism and the theory of the weak nuclear force, giving what is known as the electroweak theory. However, so far it has not proven possible to unite the strong force with the electroweak, to produce a GUT, and in fact it is not even clear that this is possible. On top of this, however, there is a fourth force, gravity, and the quest to reconcile this with the other three forces – to produce a ToE – encompasses the challenge of reconciling the theories of relativity and quantum mechanics. A synthesis of the latter two – known as quantum gravity – would form a key element of a ToE, but also necessary would be a functioning GUT, which, as mentioned, may not in fact be possible.

VERY UPSET: THE BLACK HOLE INFORMATION PARADOX

STEPHEN HAWKING'S BRILLIANT AND ELEGANT EQUATIONS DESCRIBING HAWKING RADIATION AND BLACK HOLE EVAPORATION HAD MADE HIM THE MASTER OF BLACK HOLE STUDIES IN THE COSMOLOGY COMMUNITY.

However, the reality-shaking implications of his work would push that community to breaking point, and embroil Hawking in perhaps the fiercest and most protracted dispute of his career. It all began with John Wheeler's contention that black holes have no hair.

Wheeler's "no-hair theorem" means that once information is swallowed by a black hole it is forever cut off from the rest of the universe; nothing can escape the black hole, and because

black holes have no hair they present no record of the information that they swallowed. It is as if you put a book into a safe with a one-way door, and the only record shown on the screen of the safe was the weight of the book. Crucially, the law of conservation of information (*see* box page 71) is not violated in this scenario, because although the information is now permanently inaccessible, it still exists in the universe. In our analogy, you can never read the book again but you know it is in the

THE MYSTIC PHYSICS FAN

Werner Erhard (original name John Paul Rosenberg) is an American businessman who made a fortune in the 1970s from his New Age personal development therapy programme, "est". Although est, which acquired a bad reputation for aggressive techniques and cult-like characteristics, was decidedly unscientific in inspiration and outlook, Erhard's personal enthusiasm was for theoretical physics. He converted the attic of his San Francisco mansion into a mini-conference venue and staged annual physics conferences that attracted the likes of Hawking, Leonard Susskind, Richard Feynman and Gerard 't Hooft.

Right: Businessman, controversial personal development guru and theoretical-physics fan Werner Erhard.

BLACK HOLES HAVE NO HAIR

In the world of physics, "information" can have very specific meanings; one of these refers to the variables or properties that describe a particle or energy state (also described as a wave form or quantum wave function, which relates to the fact that in quantum physics entities exist simultaneously as waves and as particles). There are a lot of these variables/ properties, and John Wheeler characterized this sort of complicated detail as "hair". When describing a black hole, Wheeler proposed, just three properties are needed: mass, angular momentum and electric charge. Black holes, he contended, have no hair. In fact, depending on what type of black hole you are discussing, you might need just one variable (see table).

Right: This image shows the quantum wave function describing an atomic system.

TYPE OF BLACK HOLE	MASS	CHARGE	MOMENTUM
Schwarzschild	Yes	No	No
Reissner-Nordström	Yes	Yes	No
Kerr	Yes	No	Yes
Kerr-Newman	Yes	Yes	Yes

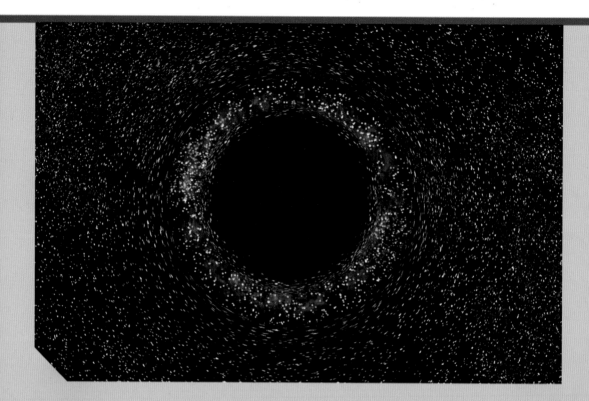

Above: Artist's conception of a black hole, with gravitational warping of the light that passes around the event horizon.

CONSERVATION THEORIES IN PHYSICS

Science, and indeed existence itself, only works because the universe is, in a fundamental sense, predictable and rule-based. A key aspect of this is that things cannot simply appear out of nowhere or vanish into nothing. These "things" might be matter, energy or information. While they might be transformed or even infinitely dispersed, they cannot vanish or appear; scientists say they are "conserved". This is the scientific equivalent of saying that two plus two must equal four; if it equals three or five, mathematics no longer works. Similarly, if fundamental aspects of nature, such as energy or information, can appear or disappear, then physics no longer works.

safe. But the theory of Hawking radiation threatened radically to transform this analogy.

In his theory of Hawking radiation, Stephen Hawking argued that the particles radiating from the event horizon are completely independent of the material entering the black hole, and if they have no relation to the original material they cannot preserve any of its information. However, at the same time, Hawking radiation is causing the black hole to evaporate. But what happens when the black hole has entirely evaporated? What has become of all the information that went into it? In our safe analogy, it is as if you removed the sides of the safe only to discover that there was nothing inside. All the information that went into it has disappeared, like a magic trick. Hawking called this the "black hole information paradox".

Hawking presented this information paradox in 1981, at one of Werner Erhard's miniature physics conferences, hosted in the spacious attic of his exclusive San Francisco mansion (*see* page 68). Probably the only person present who understood the implications of what Hawking was saying was the American physicist Leonard Susskind, who saw that if he were right, and black holes do violate conservation of information, then almost everything we know and believe about the universe would be under threat. As Hawking put it, "We wouldn't be able to predict the future [or] be sure of our past history either." Predictability and causality would be fatally undermined and quantum mechanics, hitherto perhaps the most successful and rigorously tested and proven theory in all of physics, would have turned out to be wrong. Hawking recalled that "Leonard Susskind got very upset".

Above: Leonard Susskind lecturing at Stanford University in 2013.

CONSERVATION OF INFORMATION IN QUANTUM MECHANICS

"Information" in quantum mechanics is all the properties that describe the specific state of a particle, including parameters such as spin, charge, momentum, mass, position, temperature and so on. The mathematics of quantum mechanics work whether run forwards in time or backwards in time: the calculations are always reversible, but this is only possible if the information on one side of the equation matches the information on the other side. This does not mean that a shattered glass will reassemble itself, but that the information necessary to achieve this, at least in theory, does remain after the glass is shattered. Another way of looking at it is to say that quantum mechanics predicts the probabilities of outcomes and that the probabilities of all the possible outcomes must, by definition, add up to one. An analogy would be with the probable outcomes of flipping a coin. The probability of an individual outcome will be a fraction (for example, 0.5 for heads), but the combined probability of all the outcomes must add up to one. If you summed the probabilities of all the possible outcomes and they didn't add up to one, something would have gone very wrong with the nature of reality. In quantum terms, information is equivalent to probability, and the total amount of information must "add up to one". Information cannot be lost or copied; this is known as the law of conservation of information.

ABSOLUTELY SURE HE WAS WRONG: THE BLACK HOLE WAR

HAWKING'S IMMEDIATELY INFAMOUS BLACK HOLE INFORMATION PARADOX POSED A HUGE CHALLENGE TO PHYSICISTS' UNDERSTANDING OF THE UNIVERSE, AND THREATENED THEIR ATTACHMENT TO THE THEORIES THAT UNDERPINNED THEIR WORLD VIEW.

If Hawking was right, either general relativity or quantum physics were wrong – possibly both. Leonard Susskind recalled Hawking's slightly smug smile that day in Erhard's attic, and remembered that he and his colleagues "were absolutely sure Stephen was wrong but we couldn't see why". Their quest to prove Hawking wrong touched off what Susskind

himself variously labelled *The Black Hole Wars* and *My Battle with Stephen Hawking to Make the World Safe for Quantum Mechanics*, the titles of two of his books on the topic.

One possible solution to the paradox is that the information is somehow preserved by the Hawking radiation. Remember that Hawking radiation particles are born on either side of the event horizon; the outgoing particle does not actually come from "inside" the black hole. So, in order for Hawking radiation to be carrying the information, that information would need to be a copy of the information lost inside the black hole, and copying also violates the conservation of information and thus quantum mechanics. Hawking believed that quantum mechanics was simply wrong, or at least incomplete, in the same fashion that Newton's laws of motion had been incomplete until extended by Einstein with the theory of relativity. For Stephen, the black hole information paradox was an invitation and a challenge: to construct a modified, more complete version of quantum

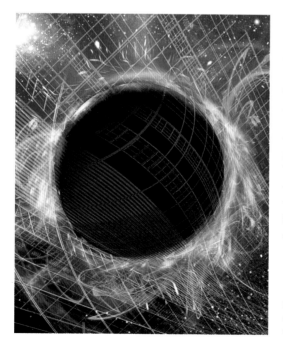

Left: Artist's concept of Hawking radiation emanating from a black hole – could this radiation resolve the information paradox?

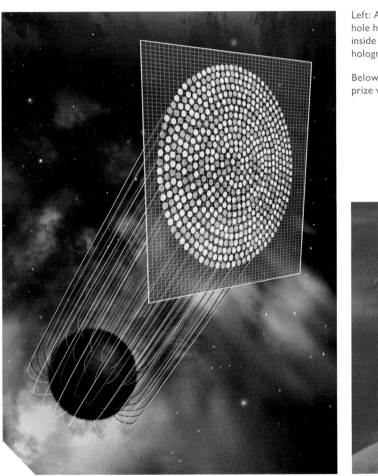

Left: Artist's visualization of the concept of black hole holography; information about the 3-D world inside the event horizon is transformed into a 2-D holograph.

Below: Dutch theoretical physicist and Nobel prize winner Gerard 't Hooft.

mechanics through developing a theory of quantum gravity.

However, Susskind and other powerful minds set out to prove Hawking wrong and rescue quantum mechanics. They wanted to show that it is relativity that must be modified. By 1992 Susskind and various colleagues had developed the theory of complementarity, which suggests an ingenious way in which information can be copied without violating the principle of conservation. Observers outside the event horizon will see information accumulate at the event horizon and then radiate back outwards in the particles of Hawking radiation, thus preserving it for the universe outside the black hole. Observers inside the event horizon will see the information that has fallen into the black hole, but because the internal and external observers cannot ever communicate, there is no paradox of duplicate information.

Complementarity in turn depends on there being a way in which information that falls

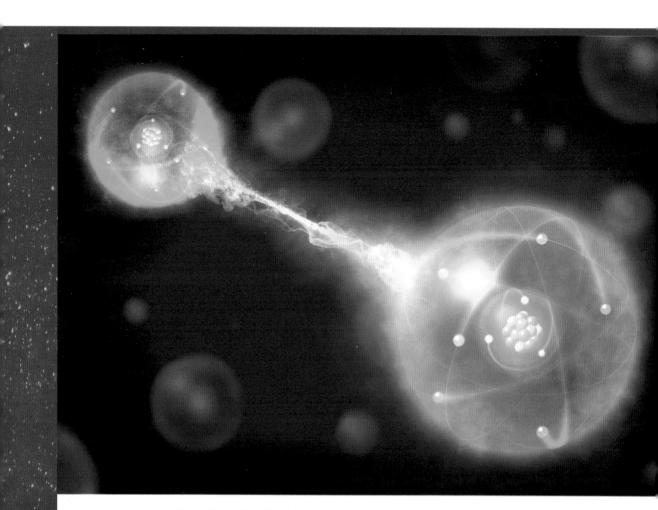

Above: When pairs of particles are quantum entangled,
determining the state of one of the pair instantly determines
the state of the other, no matter how far apart they are.

LEONARD SUSSKIND

Born in the gritty South Bronx
neighbourhood of New York, Leonard
Susskind resisted pressure from his
father to follow the family trade of
plumbing. He told his father that he
wanted to be a physicist, "like Einstein".
In 1979 he started teaching at Stanford
University in California, where he is
professor of theoretical physics at
Stanford University and director of the
Stanford Institute for Theoretical Physics.
Susskind is most noted as one of the
intellectual powerhouses behind string
theory (see box opposite).

into a black hole can be in some fashion "recorded" at the event horizon. Gerard 't Hooft developed a theory to account for this, with his holographic model. This suggests that the information of the three-dimensional world inside the black hole is transformed into a two-dimensional representation around the surface of the event horizon, in the same way that a hologram is a three-dimensional image captured on a two-dimensional surface. Susskind was able to develop a string theory validation of the holographic theory, allowing the *California Literary Review* to print the immortal headline "Susskind Quashes Hawking in Quarrel Over Quantum Quandary".

However, this was not the end of the story. The holographic model created its own paradox, after calculations suggested problematic effects of entanglement (the phenomenon whereby paired particles affect each other's states even when they are far apart) linked to black hole evaporation. By the time that half the black hole had evaporated, so much information will have been lost from the hologram at the event horizon that there is effectively no interior, and infalling information crashes into an impassable barrier at the event horizon and burns away, creating what is known as a firewall. This firewall is not supposed to exist in classical relativity, giving a paradox that threatened to destroy relativity. Susskind proposed a solution involving wormholes linking entangled particles inside and outside the black hole, but by this time Hawking had moved in a very different direction with his own work on black holes (*see* Chapter 7).

STRING THEORY

String theory is a complex theory that describes particles as vibrating loops of multi-dimensional string. In addition to the four dimensions of spacetime, string theory posits up to seven (or more) additional dimensions, although they are folded up so tiny that they are not detectable by current technology. String theory seems to unite quantum mechanics and relativity *and* provide a GUT for the Standard Model (see page 67), making it a popular candidate for a ToE, but it is incredibly hard to test or prove by experiment or observation.

Right: Conceptualization of a superstring, where up to seven or more higher dimensions are rolled up into infinitesimal vibrating strings or loops.

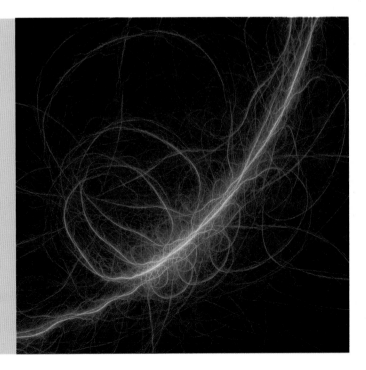

CANCELLING THE COSMOLOGICAL SINGULARITY: THE NO BOUNDARY PROPOSAL

AFTER MANY YEARS OF FOCUS ON BLACK HOLES, HAWKING'S INTEREST IN THE ORIGIN AND FATE OF THE UNIVERSE WAS REAWAKENED WHEN HE WAS INVITED TO ADDRESS A VATICAN CONFERENCE ON COSMOLOGY IN 1981.

In an audience after the conference, Pope John Paul II told him that, while he thought it was fine for cosmologists to study the evolution of the universe after the Big Bang, they should not seek to enquire into the moment of creation itself, as this was the preserve of God. Hawking later commented wryly that he was glad the pontiff did not know the topic of the talk he had just delivered, which floated for the first time the idea that perhaps there had been no moment of creation.

Over the next two years, Hawking would work intensively alongside American physicist Jim Hartle, a professor of physics at the University of California, Santa Barbara. Together they would produce a theoretical model of the early universe that they called the "no boundary proposal".

Having launched his career with proof that the universe must have started with a singularity, Hawking was now seeking a way to prove the opposite. For him, singularities were unsatisfactory: they cannot be explained by physics, as they represent a breakdown of the laws of physics. In effect, they make it impossible to answer the really big questions about creation and where the universe comes from, leaving open a considerable space for the existence of God or similar theories. By a deep exploration of the quantum gravity of the early universe, Hawking and Hartle now sought to show that a singularity is unnecessary.

Below: American physicist Jim Hartle, who collaborated with Stephen Hawking on the no boundary proposal.

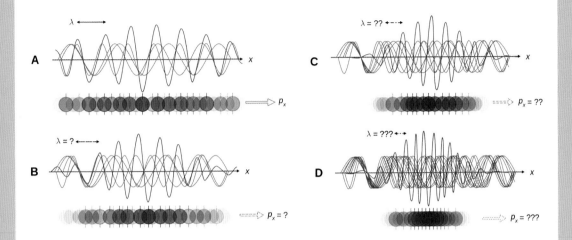

UNCERTAINTY AND SMEARING

The uncertainty principle states that it is not possible to determine precisely the position and momentum of a particle (see page 52). When trying to describe one of these properties of a particle, the uncertainty principle has what physicists called a "smearing" effect. Rather than a particle being a single, well-defined point, it is more accurate to describe it as being a kind of cloud or smear of probabilities. The probability of the particle being in one spot is greatest at the centre of the smear, and becomes smaller and smaller as you move away from this centre. A simple graphical representation of this effect is seen in a normal distribution. The distribution is heavily weighted towards the centre, but it extends out on either side. If the graph showed, for instance, the probability of a particle's momentum being of a certain value, it would indicate that the particle would have a very high probability of a momentum in the central range, but that there is nonetheless a probability, albeit vanishingly small, of its momentum being at one of the extremes.

Above: A set of illustrations showing how, as one variable of a wave function becomes more determined, the corresponding variable increases in uncertainty.

ANGRY BEES IN SHRINKING BOXES

A consequence of the uncertainty principle is that the more certain you are of one property of a particle, the less certain you can be about the other. In other words, if you precisely pin down the momentum of a particle, this greatly increases the uncertainty over its position, having the effect of smearing the position of the particle over a greater area. The more certain you are about the particle's position, on the other hand, the greater the range of uncertainty as to its momentum. For a particle this means that the less freedom it has to be in different positions, the more extreme its momentum may become. An analogy is with a bee trapped in a shrinking box. The more the box shrinks, the more excited the bee becomes and the more angrily it buzzes around. This is the quantum mechanical explanation for why, in an atom, negatively charged electrons keep their distance from the positively charged nucleus, despite the electrostatic attraction between their charges. If the electron were drawn to the proton, its location would be precisely fixed and its momentum would become practically infinite, making it too energetic to be confined at the nucleus. Hawking and Hartle would attempt to prove something similar about particles at the start of the universe.

Above: Electron orbitals in a cloud around an atomic nucleus. Quantum uncertainty explains why, despite their electrostatic attraction, the negatively charged electrons are not drawn to combine with the positively charged protons in the nucleus.

The key to Hawking's and Hartle's thinking was the uncertainty principle and its smearing effect (*see* box page 77). At the uttermost start of the universe, all matter and energy were confined in an infinitesimally small space, but like the angry bee in the shrinking box (*see* box opposite), all those particles would acquire almost infinite uncertainty in their momentum (and, by extension, their velocity). They would effectively move faster than the speed of light, breaking down the barriers between space and time. As Hawking put it, "In the very early universe, when space was very compressed, the smearing effect of the uncertainty principle can change the basic distinction between space and time." According to Hawking and Hartle, in these conditions, "we might say that time becomes fully spatialized – [...] then it is more accurate to talk, not of spacetime, but of a four-dimensional space." Spatializing time allowed them to think of the history of the universe as a curved, closed surface: a finite surface with no boundary or edge, like a ball or globe. Such a surface has no beginning or end. The classic analogy is with the North Pole of the Earth (*see* box right).

THE VIEW FROM THE NORTH POLE

If you go as far north as it is possible to go, you will reach the North Pole. But what happens if you try to travel north from there? The question is meaningless; if you move forward in any direction you will start moving south. There is no north of the North Pole; a line of longitude has no cut-off or boundary – it simply curves back around the Earth. Similarly, Hawking argues, the history of the universe has no beginning (or, presumably, end): if you went all the way back in time you would find yourself coming forward again. To ask what came before the start of the universe is to make a category error.

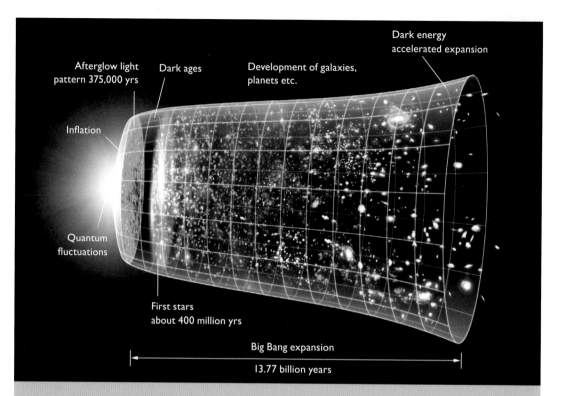

Afterglow light
pattern 375,000 yrs

Dark ages

Development of galaxies,
planets etc.

Dark energy
accelerated expansion

Inflation

Quantum
fluctuations

First stars
about 400 million yrs

Big Bang expansion

13.77 billion years

NO ROOM AT THE IN?

Crucially, the no-boundary universe removes the need for an account of creation. Questions about where the singularity came from or what came before it are no longer relevant. The universe is self-contained; it was not created, it simply is. For Hawking the proposal does not necessarily rule out the existence of God, it simply means that God had no choice about how the universe began, which would seem to strike at the heart of the concept of an omnipotent God. Perhaps more importantly from his point of view as a scientist, the no

boundary proposal shows that the entire history of the universe can be explained by science:

...there would be no singularities, and the laws of science would hold everywhere, including at the beginning of the universe. The way the universe began would be determined by the laws of science.

Above: A graphical timeline of the universe; for Hawking the important question was, what to show at the far left?

Opposite: Hawking discusses what happened at the Big Bang, at the Seattle Science Festival in 2012.

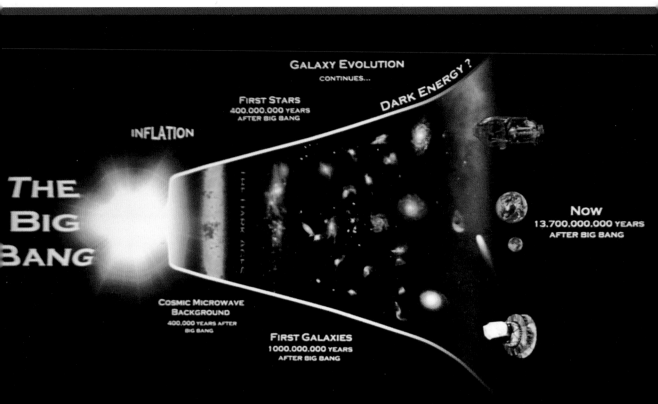

GALAXY EVOLUTION
CONTINUES...

FIRST STARS
400.000.000 YEARS
AFTER BIG BANG

DARK ENERGY ?

INFLATION

THE BIG BANG

NOW
13.700.000.000 YEARS
AFTER BIG BANG

COSMIC MICROWAVE
BACKGROUND
400.000 YEARS AFTER
BIG BANG

FIRST GALAXIES
1000.000.000 YEARS
AFTER BIG BANG

FORMATION OF
THE SOLAR SYSTEM
8.700.000.000 YEARS
AFTER BIG BANG

CHAPTER 5

A BRIEF HISTORY OF TIME

A BRIEF HISTORY OF TIME

FROM THE BIG BANG TO BLACK HOLES

STEPHEN W. HAWKING

INTRODUCTION BY CARL SAGAN

GREAT EXPECTATIONS: HAWKING'S PLANS FOR AN UNCONVENTIONAL BESTSELLER

BY THE MID-1980S STEPHEN HAWKING WAS ACADEMICALLY SUCCESSFUL AND HAD SECURED INVALUABLE FINANCIAL SUPPORT FOR HIS PERSONAL CARE. HOWEVER, EVEN THE TOP ACADEMIC POSTS IN A BRITISH INSTITUTION ARE NOT MASSIVELY LUCRATIVE AND, LIKE MANY FAMILIES, THE HAWKINGS FELT THEY NEEDED MORE THAN THEY HAD. IN PARTICULAR, THEY WANTED TO FUND PRIVATE SCHOOLING FOR THEIR DAUGHTER, LUCY.

This financial pressure was part of the motivation for Stephen to launch a bold gambit – a new work of breathtaking ambition, both intellectual and commercial. He wanted to write a book about the origins and fate of the universe and the complex science behind everything from quantum gravity and black hole entropy to string theory and the wave function of the universe. But, unusually for such a potentially complex and intellectually challenging work, he wanted it to be read by the widest possible audience among the general public. As with so much else in his life, Stephen Hawking was about to challenge seemingly impossible odds and win.

Writing or typing were not possible for Hawking; as with all other aspects of his work, he relied upon the assistance of others to interpret his almost indecipherable

Left: The fee-paying school to which the Hawkings sent their daughter, Lucy. Meeting the school fees was one of the primary motivations for Stephen to write a popular science book.

speech and translate it into words, actions or equations. When it came to writing his book, he would work closely with his graduate student, Brian Whitt. Whitt would listen to Hawking's slurred diction, ask for repetition and clarification, write down his interpretation and show it to Hawking for confirmation. This process was slow and laborious, and there was a clear incentive for Hawking to be as brief and to the point as possible. Therefore in his work he made a virtue of limiting his wordiness, thinking instead in visual and geometric fashion. Hawking would now extend this approach to his popular science book as well, as he described: "I think in pictorial terms, and my aim in the book was to describe these mental images in words, with the help of familiar analogies and a few diagrams."

AIMING FOR THE AIRPORT

By 1984, Hawking had completed a first draft of his book. However, his ambitions for the work did not seem to be matched by those of his usual publisher, Cambridge University Press, who had put out his previous, academic books. They predicted that the best he could hope for would be to sell 20,000 books a year across all markets. But Hawking had set his sights much higher. He dreamed of seeing his book on sale in airport bookshops, alongside mass-market potboilers and thrillers.

Right: The first edition of *A Brief History of Time*. The prominent photo of Hawking in his wheelchair fed accusations that the success of the book owed more to human interest than literary merit.

MAKING A MASTERPIECE

JANE HAWKING'S AMBITIONS FOR STEPHEN'S NEW BOOK WERE MODEST: SHE HOPED IT WOULD BRING IN A REGULAR IF SMALL "SUPPLEMENTARY INCOME". BUT STEPHEN MEANT BUSINESS.

He had been put in touch with a top literary agent in New York named Al Zuckerman, who knew that Stephen's extraordinary life story could prove a powerful marketing tool and make potentially juicy bait for a big publishing contract. Hawking had sent him a slim, 100-page manuscript entitled *From the Big Bang to Black Holes: A Short History of Time*, and Zuckerman began shopping it around publishers, hoping to stir up a bidding war.

One publishing house that was particularly interested in the proposal was Bantam, an imprint more usually associated with mass-market titles from bestselling authors than with challenging science books; in other words, exactly what Stephen was looking for. Peter Guzzardi, an editor at Bantam, had read a magazine story about Hawking. The report recounted the now-familiar tale of a great mind shrugging off the limitations of its fragile body to grapple with cosmic problems. Guzzardi was especially struck by the telltale detail that the soles of Hawking's shoes were pristine, because they had never touched the ground. By happy serendipity he finished reading the article en route to lunch with Zuckerman, and he was soon bidding for the rights to the book.

Above: Legendary New York literary agent Al Zuckerman, who helped broker the unprecedented deal between Hawking and mass-market publisher Bantam.

Left: By the time Hawking came to author his popular science book, his condition had progressed to the point where he was reliant on a graduate student to co-write it.

Below: Sir Isaac Newton, a previous holder of the Lucasian Chair that Hawking held, and a notable employer of analogy in conceptualizing and communicating his science.

ANALOGIES IN SCIENCE

At one point Hawking's publisher Bantam suggested using a ghostwriter to produce something more accessible, but Hawking would not have it. Early in the process of writing his book he had been warned that for every equation he included, the readership of his book would be halved. Accordingly, he put in just a single equation: Einstein's famous $E = mc^2$. In place of complex mathematics, Hawking and Whitt sought to use analogies wherever possible, painting mental images with words. In doing so, they were continuing a venerable tradition in science. Sir Isaac Newton had used analogies to illustrate concepts such as orbital and escape velocity, centrifugal force, and the effects of gravity, while analogies had been absolutely central to Albert Einstein's breakthroughs with special and general relativity. However, Hawking was also aware of the limitations and dangers of analogy; by definition, analogies are inexact and can be misleading if extended too far.

Above: One of the most contentious areas of public debate about Hawking, pictured above in characteristic good humour, is over his attitude to religion. His often ambiguous statements on the topic have meant that both atheists and religionists co-opt him to their side.

Bantam offered $250,000 for US rights, with another £30,000 from Bantam-Transworld for the UK rights. These were unprecedented numbers for a book about science. When Hawking travelled to the United States for a meeting in Chicago, Guzzardi took the opportunity to meet him in person. Once he caught up with the racing electric wheelchair, he introduced himself. Hawking, speaking through Brian Whitt, wasted no time on small talk, demanding to know if Guzzardi had brought the contract. The terms were agreeable and Guzzardi, an editor with no science background whatsoever, found himself editing a manuscript about black hole topology and spatialized time dimensions. His input would be key to transforming Hawking's dense and technical text into something that could be put before the general public.

The process of producing a second draft of the book proved to be lengthy and difficult.

Hawking refused to compromise on technical accuracy, and perhaps struggled to put himself in the shoes of those less intelligent than himself. Guzzardi's role was to represent the lay person: if he could not understand what he was reading, he would send it back to Hawking with questions and requests for explanation, and would keep on doing it until he could.

In the end it would take nearly two years to produce a final draft of the book, partly thanks to Stephen's brush with death, traumatic operation and subsequent recovery, and search for a new way to communicate (*see* pages 92–3). Hawking was able to show a late draft to his father – not long before Frank's death in 1986 – but it was not until around the summer of 1987 that a final manuscript was delivered to the publishers. A notable change was to the title, to which had now been added the word "brief"; Guzzardi over-ruled Hawking's late misgivings about the title, insisting that he liked the touch of whimsy it brought.

THE LAST WORDS

There was also editorial debate over the controversial final passage of *Brief History*, in which Hawking suggests that a Theory of Everything would enable humanity to "know the mind of God". In his biography, *My Brief History*, he would later reflect: "In the proof stage I nearly cut the last sentence in the book.... Had I done so, the sales might have been halved." In the wake of the book's enormous popularity, this passage has been touted as evidence both for Hawking's atheism and for suggestions that he was a secret believer in God.

NEAR-DEATH CRISIS

IN THE SUMMER OF 1985 STEPHEN WAS PLANNING TO SPEND A MONTH AT THE EUROPEAN ORGANIZATION FOR NUCLEAR RESEARCH (CERN) IN GENEVA. HE WOULD BE ACCOMPANIED ON HIS TRIP BY HIS NEW ASSISTANT LAURA GENTRY, ALONG WITH SOME OF HIS NURSES AND STUDENTS.

As had happened in previous years, Jane Hawking took the chance to give the children a camping holiday, taking a road trip across Europe en route to meeting up with Stephen.

This time she went with Jonathan (*see* page 98). Jane, Jonathan, Lucy and Tim were due to meet Stephen at Bayreuth, the German town in which a Wagner festival is held each year. The

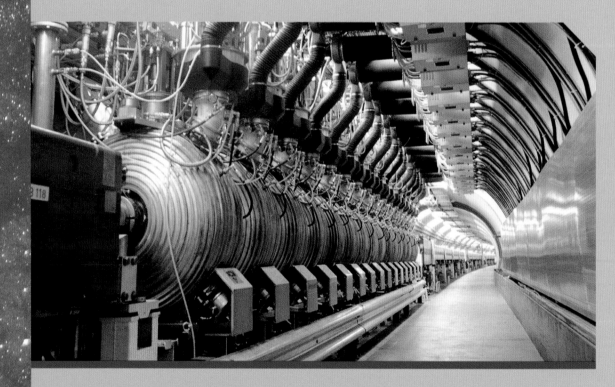

Above: The super-proton-synchroton (SPS) accelerator at CERN in Geneva, the high-energy physics institution that Hawking visited often in his life.

night before they were due to arrive, Jane called from a payphone to finalize arrangements, only to have a bombshell dropped on her. A panicked Laura Gentry informed her that Stephen had been taken perilously ill in Geneva, stricken with pneumonia. He was in hospital.

At the hospital, Jane was faced with a grave predicament. Stephen was in an induced coma, on life support with a machine to breathe for him. His only option for survival was to have a tracheotomy, an operation that involved cutting a hole in his throat so that he would be able to breathe without obstruction, relieving the constant risk of coughing and choking that resulted from degeneration of his throat muscles. However, a tracheotomy would exact a heavy price, since it would mean Hawking could never speak again, finally losing the limited speech of which he was still capable. The doctors suggested that it might be kinder simply to switch off the life support.

Jane was unequivocal: Stephen would survive, whatever it took. She had him flown back to Cambridge by air ambulance. The tracheotomy was performed and Stephen started on the slow road to recovery. Henceforth, he would breathe through a small opening in his throat at collar level, and would require additional care to keep the airway clear and suction out fluids whenever necessary.

THE GENIUS GRANT

Jane Hawking was insistent that Stephen would come back home rather than go to a nursing home, but she was faced with the daunting cost of round-the-clock care. It was Kip Thorne who came to the rescue, putting her in touch with the John D. and Catherine T. MacArthur Foundation, a charitable body that gives what are sometimes called "genius grants" to support elite creative thinkers. Securing a MacArthur grant to pay for nursing meant that Stephen was able to return home in November 1985, after three months in hospital.

Above: Kip Thorne, pictured in 2016. One of Stephen's staunchest friends, Thorne helped set up the MacArthur Foundation grant that would help meet Stephen's medical costs.

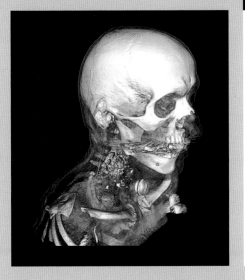

Above: A CT scan of a person who has had a tracheotomy, with a ring inserted into the opening to hold it open. This can aid breathing when upper airways are obstructed or not functioning.

SPEAKING WITHOUT WORDS

STEPHEN HAWKING HAD ALREADY OVERCOME APPARENTLY INSURMOUNTABLE CHALLENGES THAT WOULD HAVE DERAILED MANY PEOPLE – NOW HE FACED YET ANOTHER. HOWEVER, AS WITH OTHER ASPECTS OF HIS DISABILITY, HAWKING WOULD TURN A CHALLENGE INTO AN OPPORTUNITY.

Deterioration of his neuromuscular abilities meant that he had already almost lost the power of speech. Only those closest to him – family, carers and graduate students – were able to interpret his increasingly indistinct words, and then only with difficulty. There

are strange echoes here of "Hawkingese", the idiosyncratic "dialect" of his youth (*see* page 11).

As Hawking embarked on his recovery, it seemed initially as though the language barrier had grown even higher. At first he could only communicate by raising an eyebrow to confirm a choice of letter on a transparent perspex letter board that would be held up by his interlocutor. It was grindingly slow and Hawking was profoundly discouraged. He later said that this was the only occasion on which he contemplated suicide – and even that he tried to stop breathing but could not overcome his automatic response.

Things improved markedly when the Hawkings were sent Equalizer, a computer program devised by the Californian computer engineer Walt Woltosz, in order to help his disabled mother-in-law. It allowed the user to choose words and letters on a computer screen using a simple device (Hawking's students rigged up a computer-mouse-like trigger he

Left: Walt Woltosz, the founder of Words+, with the PegasusLITE, a battery-operated voice synthesizer and descendant of the system that Hawking adopted.

Below left: The tough, knock-resistant housing for Hawking's computer and speech synthesizer, enabling it to be more easily transported.

Below right: The interface for Hawking's wheelchair-mounted, portable voice synthesizer system, showing how an auto-complete function helped speed communication by suggesting lists of possible matches.

could operate with his limited manual ability), and employed a simple speech synthesizer to convert them into spoken words. Brian Whitt recalled that he visited Hawking on the first occasion of him using the program. Stephen was direct as usual, typing "Hello", followed by "Will you help me finish my book?". Although at best he could manage about 15 words a minute, Hawking would later acknowledge that, once he had mastered the new system, he was now better able to communicate than before his tracheotomy.

CHOOSING A VOICE

Hawking's synthesized voice was to become his signature sound (he even copyrighted it), but of course he could have chosen to change it or to vary it on a regular basis. He proclaimed himself happy with the voice originally supplied to him because, although devoid of emotion, it included enough intonation to avoid being wholly robotic. In later years, he declined offers to "upgrade", pointing out that he, like others, had grown accustomed to the voice, such that changing it would feel like changing his character. Hawking himself described the voice as having an American accent, although different people ascribe different identities to it, from Scandinavian to South Asian.

SUCCESS OF THE BOOK

EQUIPPED WITH HIS NEW METHOD OF COMMUNICATION, AND WITH THE ASSISTANCE OF BRIAN WHITT, HAWKING WAS ABLE TO GET BACK TO WORK ON HIS BOOK, PRODUCING A FINAL DRAFT BY THE SUMMER OF 1987.

A Brief History of Time: From the Big Bang to Black Holes was launched in America on 1 April 1988, and in the UK on 16 June, becoming a publishing phenomenon as it sat at the top of the bestseller lists for weeks, months and eventually years. Hawking achieved his dream of seeing it on the shelves of airport bookshops, and it was translated into many languages.

What were the secrets of the book's success? People responded to its great ambition and enjoyed its cosmic scope, while the hard work of Guzzardi meant that, although still dense and challenging, the text could be followed by intelligent lay people. It was also leavened by Hawking's sense of humour. But inevitably the marketing angle played its part, with the US edition in particular featuring a photo of Hawking in his wheelchair, juxtaposed against a background of stars. There was widespread sneering that buyers simply got caught up in the hype, that virtually no one actually read *Brief History* (*see* box below), and that many buyers bought the book simply in order to look clever.

THE MOST UNREAD BOOK IN HISTORY?

Could *Brief History* be the book of which the most copies have been sold but not read, or at least not finished? This suggestion has even spawned a "Hawking index" that measures exactly this (*Brief History* scores very highly, alongside E. L. James's *Fifty Shades of Grey*). There is anecdotal evidence that few readers get past Chapter 4, while Hawking himself has written that he regrets not making more effort to explain concepts such as imaginary time. He also acknowledged that many copies were purchased simply to decorate coffee tables, but he contended that this put him in good company, alongside other widely owned but little-read books such as the Bible and the works of Shakespeare.

Above: Various foreign-language editions of Hawking's bestseller, alongside the UK first edition. In the US and UK markets, the book sold so relentlessly that the publishers delayed putting out a paperback version for an unprecedented length of time.

RECORD BREAKER

A Brief History of Time has sold over 10 million copies, and been translated into more than 40 languages. It was on the bestseller list of *The New York Times* for 147 weeks and that of the British newspaper *The Sunday Times* for an unequalled five years (247 weeks).

CHAPTER 6
LONELY AT THE TOP

AN UNCONVENTIONAL ARRANGEMENT

WHEN JANE MARRIED STEPHEN SHE HAD DELIBERATELY LOOKED AWAY FROM THE REALITIES OF HIS MEDICAL PROGNOSIS AND WHAT IT MIGHT MEAN FOR THEIR LIVES TOGETHER.

As his disability worsened, she worked harder and harder to provide the practical and often intimate care that he needed while at the same time juggling multiple other roles: caring mother; supportive academic wife; scholar; personal assistant; travel agent; secretary; and amanuensis to a figure of global stature with diverse and mounting professional and public duties. The physical and emotional stress she suffered is hard to imagine. At the same time she had to deal with personal challenges around faith, self-worth and identity. Jane's

dogged pursuit of her PhD, for instance, stemmed from her sense that an academic wife at Cambridge must somehow demonstrate her own intellectual bona fides. Her Christian faith was often at odds with and sorely tested by Stephen's own beliefs (*see* box opposite).

One of Jane's primary outlets for emotional respite came from joining the choir at nearby St Mark's church, in late 1977. The organist there was Jonathan Hellyer Jones, a recently widowed and lonely young man with whom Jane had much in common, including a love of

Opposite: Jane and Stephen Hawking in Paris, in 1989. Success only served to increase the pressures on Jane as she struggled to fulfil multiple roles.

Right: The Hawkings at home with their young family in Cambridge, 1981. The unique challenges they faced required unconventional solutions.

music and a quiet Anglican faith. He became a regular visitor at the Hawking household, teaching Lucy the piano and volunteering to help provide physical care to Stephen. Jane and Jonathan fell in love but their relationship remained platonic. Stephen accepted this development, demanding only that Jane continued to love him, and indeed the couple had their third child together, Timothy, shortly after this. The household settled into an unconventional ménage à trois, but health crises for Stephen in 1980 and again in 1985 would bring major changes.

When Stephen's condition deteriorated to the point where the couple were forced to admit nurses and carers into their house and their lives, the change proved to be a double-edged sword. On the one hand, Jane was freed from some of the impossible demands on her, but on the other, her diligently maintained domestic sphere was now turned upside down by multiple outsiders.

NOTHING PERSONAL

Hawking claimed not to be an atheist, but for an agnostic he was rather extreme. In his writings and pronouncements Stephen often referred to God, but his use of the term was imprecise, encompassing both God as a metaphor for the laws of physics and as a catch-all term for concepts beyond the reach of – or relevance to – science. One thing he was very definite about was that he did not believe in a personal god, or in one with any agency, or indeed interest, in the universe. This put him squarely at odds with his wife, Jane, whose strong and abiding personal faith had been a source of strength to her in dealing with the many challenges of their marriage and life together. At times she found his lack of belief hurtful.

Left: Jane with Jonathan Hellyer Jones, who came into the Hawkings' lives in 1977.

SEPARATION AND DIVORCE

AFTER HIS HEALTH CRISIS OF 1985, STEPHEN REQUIRED ROUND-THE-CLOCK NURSING CARE.

Above: Hawking with his second wife Elaine, attending a film premiere in 2004.

One of the nurses recruited was Elaine Mason, a strong person and even stronger character, with striking red hair and an earthy attitude. An experienced and capable nurse, she soon hit it off with Stephen and they enjoyed each other's sense of humour. Her husband, David, was a computer engineer who helped develop the speech technology upon which Stephen relied. By adapting a small computer with a synthesizer unit, David created a portable speech system that could be attached to Stephen's wheelchair, allowing him to take his voice with him wherever he went.

With round-the-clock care, and nurses and other support staff coming and going, the Hawking household was almost a public space. To add to strains on the Hawking marriage, Stephen began to favour Elaine Mason more and more. She it was who would most frequently accompany him on his many trips, and her protectiveness and forcefulness caused tensions in the world of the Hawkings, including with his co-workers and even family members.

The runaway success of *Brief History* had raised Stephen's already high profile to stratospheric levels. He featured in newspapers and magazines around the world, and was fêted as the "Master of the Universe" by both *Newsweek* magazine and a television special about him. He was deluged with awards and

Right: Elaine and Stephen on their wedding day; they had a ceremony at a register office followed by a church blessing.

honours. Jane felt compelled to maintain a positive front, telling one interviewer of her "sense of fulfilment that we have been able to remain a united family". But the truth was that in the next few years they would grow ever further apart.

Stephen basked in his celebrity status and was determined to exploit to the fullest all the opportunities that came his way, while also taking on himself the burden of responsibility that came with being a global role model for science and the disabled. Jane increasingly felt that she and the family could not keep up and that Stephen was content to leave them behind, travelling the world in the company of Elaine. Meanwhile, Jane had her relationship with Jonathan and preferred to stay out of the spotlight to pursue passions such as teaching, gardening, singing and religion. In the summer of 1990, Stephen told Jane that he was leaving her for Elaine, and they moved out of the West Road home to move in together elsewhere. In 1995 Stephen and Jane were divorced and he married Elaine. Two years later, Jane and Jonathan were married. In 2006, after 11 years of marriage, Stephen and Elaine were divorced amidst rumours of physical and emotional abuse.

ABUSE ALLEGATIONS

In 2000 and again in 2003 the police investigated claims made by Hawking's carers and family that he was being abused. The finger was pointed at Elaine Mason. Unnamed sources made accusations through the newspapers of incidents of abuse ranging from bruising and cuts to humiliating treatment, deliberate submerging of his tracheal opening and leaving him out in the sun until he got heatstroke. Stephen Hawking himself always vehemently denied all the claims and the police dropped their investigations.

CHAPTER 7

CELEBRITY SCIENTIST

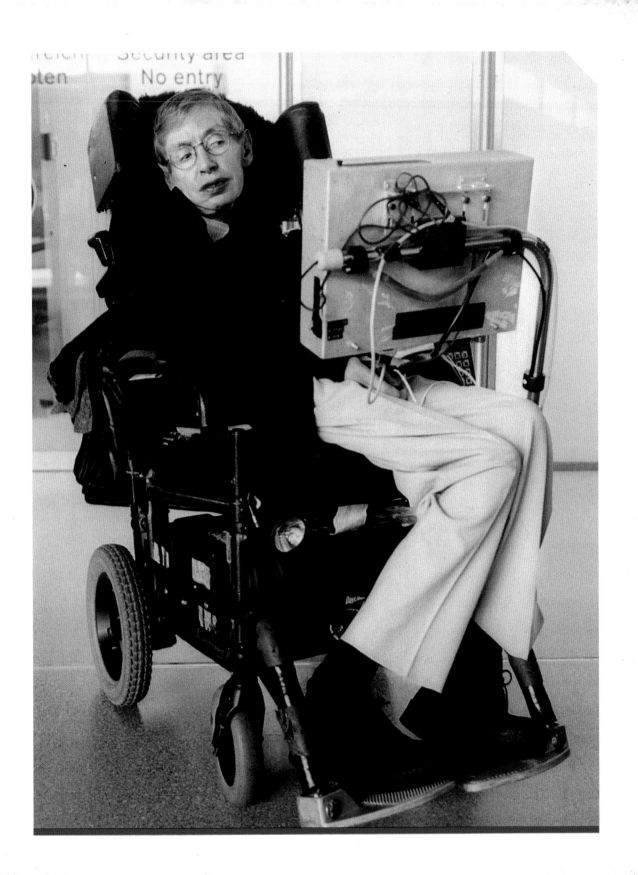

FLYING HIGH

THE MASSIVE SUCCESS OF *BRIEF HISTORY*, COUPLED WITH HAWKING'S SCIENTIFIC AND PERSONAL ACHIEVEMENTS, OPENED UP A NEW WORLD FOR HIM: THAT OF A JET-SETTING INTERNATIONAL CELEBRITY. HE TOOK FULL ADVANTAGE.

From 1989 until the end of his life he kept up an almost unceasing and hectic schedule of appearances, visits and projects around the world.

Awards and recognition showered upon him. Highlights included being made a Companion of Honour in 1989 – a royal honour bestowed upon just 65 Britons at a time – although Hawking would later decline a knighthood for political reasons. The previous year he had been in Jerusalem to accept the Wolf Prize, a prestigious scientific accolade second in stature

Opposite: President Bill Clinton and Hawking at a Millennium Evening event at the White House, with Hawking's cameo from *Star Trek: The Next Generation* pictured on-screen.

Left: Hawking in the 1991 Errol Morris film adaptation of *Brief History*. Now he was a film star as well as a bestselling author and prize-winning scientist.

only to the Nobel, while the following year he would be serenaded by Ella Fitzgerald at a ceremony at Harvard University, in which both were awarded honorary degrees. Hawking's was added to those he had already received from both Oxford and Cambridge universities, among many others.

Also in 1990, Hawking was approached by Steven Spielberg, who wanted to produce a documentary film of *Brief History*; it would go on to be directed by acclaimed filmmaker Errol Morris and win multiple awards. As Hawking's profile continued to rise, his talks and lectures attracted ever-greater crowds. People queued for hours for a seat in the lecture hall when he gave a lecture at Berkeley near San Francisco in 1993, while in 1995 he filled the Royal Albert Hall in London at an event to raise money for an ALS charity. In 1998, he was invited by President Bill Clinton to give one of the lectures in the White House Millennium Evening lecture series.

Hawking's international travel was constant and varied. Japan was a favourite country –

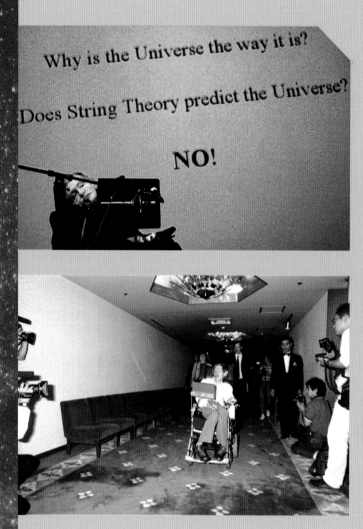

Why is the Universe the way it is?

Does String Theory predict the Universe?

NO!

Left: Hawking addresses a symposium of cosmologists at Stockholm University, 19 August 2003. Before the symposium, he received the Oskar Klein Medal.

he would visit seven times, with the famous bullet train having precious minutes added to its timetabled stopping time to accommodate his wheelchair disembarkations. His travels ranged from Chile and India to South Korea and even Antarctica, which he visited with Kip Thorne and other scientists in 1997. After 2000 he mostly travelled by private jet, and the pace of his voyaging was relentless. In 2003, for instance: Hawking spent a month at the Mitchell Institute for Fundamental Physics at Texas A&M; attended a meeting on cosmic inflation at the University of California-Davis; received the Oskar Klein Medal from the Royal Swedish Academy of Sciences; attended a Nobel symposium in Sweden; spent two months at Caltech; and then spent time at Case Western Reserve University in Cleveland, Ohio. In 2005 he visited eight different locations/institutions, including receiving the James Smithson Bicentennial Medal from the Smithsonian Institute in Washington.

Above: Hawking in Tokyo in 1990. He visited Japan frequently, although his travels took him to every continent of the world.

ZERO GRAVITY FLIGHT

One of Hawking's most famous exploits was to experience four minutes of weightlessness on a zero gravity flight in 2007. The Florida-based Zero Gravity Corporation took him up in one of their modified jets, lifting off from the Kennedy Space Center for several hours of swooping and diving, as part of Hawking's drive to drum up interest in space exploration (see pages 146–7). He was also intent on demonstrating, as he had done for so much of his life, that disability does not have to be a barrier to adventure.

Above: Stephen Hawking in freefall, on board a plane that dives to give passengers the experience of weightlessness, bringing to life one of Einstein's most famous thought experiments.

PROBLEM GAMBLER: THE FURTHER SCIENTIFIC WAGERS OF STEPHEN HAWKING

HAWKING HAD ENJOYED HIS CYGNUS X-1 WAGER WITH KIP THORNE IN THE 1970S (*SEE* PAGE 58), AND WENT ON TO MAKE A NUMBER OF OTHER SCIENTIFIC WAGERS.

Owing to his status as a celebrity scientist, these achieved a remarkably high profile, on occasion making newspaper headlines. His most notable wagers were: his 1991 bet against Thorne and fellow Caltech theoretical physicist

John Preskill, over the existence of naked singularities; his 1997 bet, in which he teamed up with Thorne to oppose Preskill on the issue of black hole information loss (*see* pages 114–5); his 2000 bet with Gordon Kane over the existence of the Higgs boson; and his 2001 wager with South African physicist Neil Turok that primordial gravitational waves would be detected and would prove that the universe inflated into existence rather than being part of a cyclic continuum.

Hawking's first bet with Preskill and Thorne stemmed from an argument about whether or not a singularity – the point of infinite density at the heart of a black hole, where the laws of physics break down – must always be "clothed" with an event horizon. Hawking agreed with Roger Penrose, who had formulated a "cosmic censorship" rule, which stated that the reality-bending, law-breaking mischief of which a singularity is capable must be decently hidden from the rest of the universe behind the veil of an event horizon. If there is no way for information or anything else to get from

Left: John Preskill, theoretical physicist and frequent sparring partner of Hawking in scientific wagers.

Opposite above: Hawking and Neil Turok, who would bet on the existence of primordial gravitational waves.

Opposite below: Peter Higgs and Hawking. Their disagreements generated headlines.

FEUDING GIANTS

In 1996 Hawking published a paper predicting that the Higgs boson could not be found, and in 2000 when the Large Electron Positron (LEP) collider at CERN reached the limits of its power without having detected the boson, he collected $100 from Kane. Peter Higgs did not appreciate this, and in 2002 was reported to have grumbled about it to fellow physicists at a dinner, complaining: "It is very difficult to engage [Hawking] in discussion, and so he has got away with pronouncements in a way that other people would not. His celebrity status gives him instant credibility that others do not have."

the singularity to outside observers, it need not matter so much in physical terms if the singularity is messing with cause-and-effect or allowing time travel, since these will have no effects on the wider universe. Preskill and Thorne hoped it might be possible for naked singularities to exist, pointing out that if this were the case, and one could be observed, it would provide incredible insights into the realm of quantum gravity.

In 1991 Hawking, declaring that "naked singularities are an anathema and should be prohibited by the laws of classical physics", bet Thorne and Preskill that such singularities cannot, even theoretically, exist. The loser would pay up $100 and "reward the winner with clothing to cover the winner's nakedness", said clothing to be decorated with "a suitable concessionary message".

In 1997, to Hawking's dismay, a computer simulation by Matthew Choptuik of the University of Texas in Austin showed that, in certain very restrictive hypothetical circumstances, a naked singularity might be possible. Hawking refused to pay over the $100, and although he did present Preskill and Thorne with T-shirts, they bore the legend "nature abhors a naked singularity" imprinted over a picture of a naked lady. Hawking then drew up a revised bet, re-worded to rule out

Above: Conceptual art imagining time travel through a wormhole; the kind of paradox-producing phenomenon that dismayed Hawking.

Opposite: A view of the LHC, superimposed on to which is a simulation of a particle-scattering collision event.

THE GOD PARTICLE

The Higgs boson, nicknamed the "God particle", is the force carrier (the particle that mediates a particular force interaction) for the Higgs field. The Higgs field is a field predicted by Scottish physicist Peter Higgs, and others, which gives mass to other particles, and which was originally formulated to explain inconsistencies or puzzles in the Standard Model of fundamental particles and forces. Higgs predicted the field back in the 1960s, but proving its existence would be a monumental challenge because it was clear that the boson that mediates the field must have a very high mass/energy (remember that thanks to $E=mc^2$, mass and energy are equivalent, so particle physicists discuss mass in terms of energy levels). To detect new particles, scientists smash together other particles and look at the debris generated by the collisions. In order to detect such a high-mass boson, it would be necessary to generate colossally energetic collisions. Successively larger particle accelerators from the 1970s to the 2000s were hoped to be able to achieve these energies.

Below: Hawking during a visit to CERN in 2006. He often visited to work with theoretical physicists based there.

special cases; this version has never been settled, although work done in 2017 suggested that Hawking and Penrose might have been right after all.

In 2000 Hawking made a bet with Gordon Kane, professor of physics at the University of Michigan, over whether the Higgs boson (*see* boxes pages 109 and 111) would be discovered. The background to this bet was an unseemly feud between Hawking and

Peter Higgs, two of the grand old men of British physics, that revealed much about the attitude of Stephen's peers towards him and his celebrity status (*see* box page 109).

Hawking was upset by Higgs's comments, and although the two smoothed things over in private, tension flared again as CERN's Large Hadron Collider (LHC), the most powerful particle accelerator yet, neared switch-on in 2008. Hawking had made another $100 bet with Kane on whether the LHC would detect the Higgs boson, and declared loudly through the media that it would be much more exciting for physics if it were not found. Again, Higgs publicly disagreed and the press made much of this feud at the highest levels of the scientific establishment. In fact the LHC did detect the Higgs boson, with the discovery announced in 2012, and Hawking paid over the $100 to Kane. Higgs was able to savour both the satisfaction of vindication and, the following year, the Nobel Prize in Physics, an accolade that always eluded Hawking (*see* page 153).

BETTING ON COSMIC RIPPLES

Stephen Hawking explored in his work many aspects of the theory of inflation, a widely accepted explanation for how our universe has come to look as it does, having started off in a Big Bang. Inflationary theory says that almost immediately after the Big Bang, spacetime inflated at incredible speed to give the mostly smooth, uniform universe we see today. Hawking and others have predicted that evidence of cosmic inflation will be encoded in the cosmic microwave background (CMB) radiation, in the shape of ripples or fluctuations left by gravitational waves generated by the Big Bang. Detection of these ripples would be evidence for the primordial gravitational waves and thus proof of cosmic inflation. South African physicist Neil Turok has been one of the few to dissent from the cosmic inflation theory, suggesting that there might be alternative explanations. As an example he developed a theory that the universe is cyclic, and this theory predicts that there will not be ripples in the CMB. In 2001 Hawking and Turok made a wager over whether gravitational waves would be detected. In 2014 a team looking for these ripples made global headlines by declaring that they had been detected. Hawking gleefully proclaimed to the global media that he had won his bet and told Turok that he ought to hand over $200. Turok wisely advised caution, and Hawking was left with egg on his face when it was admitted that the ripples detected were actually artefacts caused by galactic dust. The bet remained unsettled.

Above: Illustration showing cyclic expansion and contraction of the universe.

THE INFORMATION LOSS REVERSAL

BY 2003 HAWKING'S BATTLE WITH SUSSKIND AND OTHERS, OVER THE FATE OF INFORMATION THAT FALLS INTO A BLACK HOLE, HAD BEEN RUMBLING ON FOR OVER 20 YEARS.

At a conference that year to celebrate the development of further support for his holographic theory, Susskind compared Hawking to a soldier lost in the jungle who doesn't realize the war is over. Hawking had time to consider his response while he was in hospital recovering from another life-threatening bout of pneumonia. By the following year he was ready to make a pronouncement, widely heralded as likely to be a public about-face.

Above: At the 2004 Dublin conference, John Preskill holds aloft the baseball encyclopaedia presented to him by Hawking in conceding their bet over information loss in black holes.

SUMMING OVER HISTORIES

Richard Feynman's "sum over histories" or "sum over paths" approach describes how events in the real world are represented in terms of quantum mechanics. Because quantum physics is probabilistic, in quantum mechanical terms a particle travelling from A to B can take a potentially infinite number of paths between the two points, although most of these will be fantastically unlikely. The actual path can be determined by adding up (summing) all the probabilities. When this is done most of them interfere with one another – that is, cancel each other out – and only the actual path is left. For "path" we can substitute "sequence of events" or "history", hence this approach is also known as summing over histories. Hawking sought to apply this method to the many different possible histories of a black hole, and this same concept would later reappear in his final work on cosmic history (see page 128, "Top-down cosmology").

Right: Quantum physics says that, when describing how a particle goes from A to B, it is necessary to sum all the possible paths it could have taken.

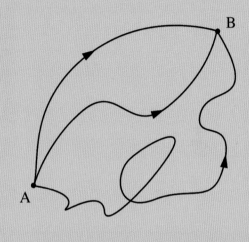

With characteristic showmanship, Hawking let it be known that his proclamation would be forthcoming at a conference in Dublin, and the global media descended on the venue to the bemusement of many scientists. Hawking took the stage and announced that he had indeed decided that information is not lost when it falls into a black hole, but he confounded many by coming up with an entirely new and obscure explanation for his conclusion. Drawing on Richard Feynman's theory of summing over histories (*see* box above), Hawking suggested that while information may be lost in our particular universe, it is preserved overall when factoring in all alternative histories, including those in which a black hole never formed.

He then went on to make a stagey concession to John Preskill over their 1997 bet about whether information is lost when it falls into a black hole. Hawking had Preskill presented with a baseball encyclopaedia to represent the information at issue, which Preskill duly brandished triumphantly for the cameras, as though, he later said, he had won Wimbledon and was showing off the trophy. For many at the conference, who found Hawking's explanation unconvincing and short on detail, the affair was puzzling at best, and at worst a stunt in which Hawking hijacked the conference to boost his media profile.

A PARTY FOR TIME TRAVELLERS

HAWKING'S ROLE AS A SCIENCE COMMUNICATOR AND POPULARIZER BECAME INCREASINGLY IMPORTANT TO HIM, HENCE HIS FONDNESS FOR GRAND AND PUBLIC GESTURES THAT OTHERS MIGHT DESCRIBE AS STUNTS. A NOTABLE EXAMPLE WAS A PARTY THAT HE HELD IN 2009, TO WHICH PEOPLE WERE ONLY INVITED A YEAR LATER. HAWKING HELD THE PARTY TO PROVE A POINT ABOUT TIME TRAVEL.

One of the challenges to conventional physics offered by singularities in the fabric of space and time is that they might make it possible to travel back in time. Recall that in the rubber-sheet model of spacetime, a singularity is like a bottomless well. What if two of these bottomless wells were to join together? This would create a tunnel or passage from one region of spacetime to another, known as a wormhole. Hawking was particularly exercised by a suggestion from Kip Thorne, who had calculated that the throat of a wormhole

Above: Conceptual graphic of a wormhole, a phenomenon that Kip Thorne suggested could be used to travel backwards in time.

could be held open with the help of exotic matter, and that this would allow someone to travel through it, and thus through space and time. If time travel really is possible, it creates numerous paradoxes (*see* box) and threatens to violate causality and conservation principles.

Hawking suggested what he called the chronological protection hypothesis: the universe somehow prevents time travel from occurring in order to prevent such paradoxes. If he were wrong, time travellers from any point in history after the invention of time travel would be able to travel back in time and attend his party; it wouldn't matter that the party had only been made public after the event. On 28 June 2009, Hawking prepared a room with balloons, nibbles and a banner proclaiming "Welcome, Time Travellers", but, he lamented, "no-one came", even though he subsequently presented an invitation on his 2010 television series *Into the Universe with Stephen Hawking*. "You are cordially invited to a reception for Time Travellers", it read, giving the date and co-ordinates of the party. The fact that no one showed up, Hawking said, proved that time travel will never be invented because it is indeed impossible.

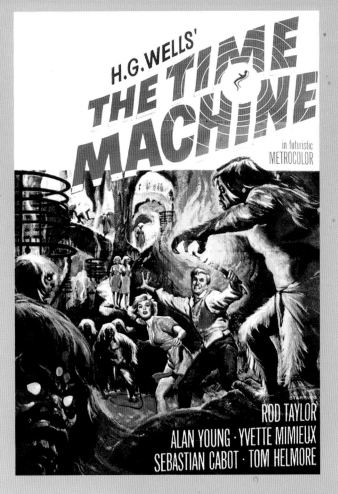

Above: As H. G. Wells' classic *The Time Machine* suggests, time travel would be bound to cause unpleasant complications.

TIME TRAVEL PARADOXES

One of the most famous time travel paradoxes is the grandfather paradox: a time traveller could go back in time and kill her grandfather, which would mean she would never have been born, which would mean she could not have travelled back in time in the first place. Another one is called the ontological paradox: what if you learned the secret of time travel when your future self journeyed back in time to tell you, and used it to build a time machine to go back in time and tell yourself the secret of time travel? Where did the idea come from in the first place?

CHAPTER 8

THE GRAND DESIGN

PROPHET OF DOOM

AS PROBABLY THE MOST FAMOUS SCIENTIST IN THE WORLD, STEPHEN HAWKING CAME TO BE LOOKED UPON AS AN ALL-ENCOMPASSING SOURCE OF WISDOM, AND IN PARTICULAR FOR ORACULAR PRONOUNCEMENTS ABOUT SCIENCE, TECHNOLOGY AND THE FUTURE. ONE OF HIS MOST PERSISTENT THEMES IN THIS ROLE WAS OF GATHERING THREATS TO THE SURVIVAL OF CIVILIZATION AND OF HUMANITY AS A SPECIES.

One of Hawking's first high-profile warnings came in 1998, during his White House Millennium Evening lecture, when he spoke about the risks of over-population and over-consumption, and talked darkly of the dangers of malign genetic engineering. However, Hawking had been sounding Doomsday alarms since at least 1994, when he addressed a Macworld Expo in Boston about the threat posed by computer viruses – a new form of life that we had created ourselves in Frankenstein-like fashion. This conference was also notable as being an early occasion in which Hawking floated one of his solutions for terrestrial limitation to human survivability (*see* box).

VON NEUMANN PROBES

Pointing out that human lifespans are too short for interstellar travel, Hawking touted the concept of self-replicating space probes (known as von Neumann probes, after the Hungarian-American physicist John von Neumann, who first suggested them), which could survive long journeys, and build and despatch more probes if they found suitable sites. The need for humanity to find some way to colonize extraterrestrial locations, and thus reduce the risks attendant on being a single-planet species, would become a constant in his pronouncements and lead to his involvement in some bold schemes (see page 147).

Right: John von Neumann, who first suggested that self-replicating robot probes are the best option for interstellar exploration.

The new millennium brought an accelerating stream of Doomsday warnings from Hawking. In 2001, in the wake of 9/11, he spoke again of the dangers of artificial lifeforms (at this point he still had viruses in mind) and the need for humanity to colonize space. By 2007, Hawking and his friend Sir Martin Rees – by now the Astronomer Royal – were appearing at the Royal Society in a dramatic illustration of their fears for the future. There they presided over the moving forward of the hands of the "Doomsday Clock", the figurative device instituted by the Bulletin of the Atomic Scientists to illustrate the current threat level of civilization-ending apocalypse (midnight represents the stroke of doom).

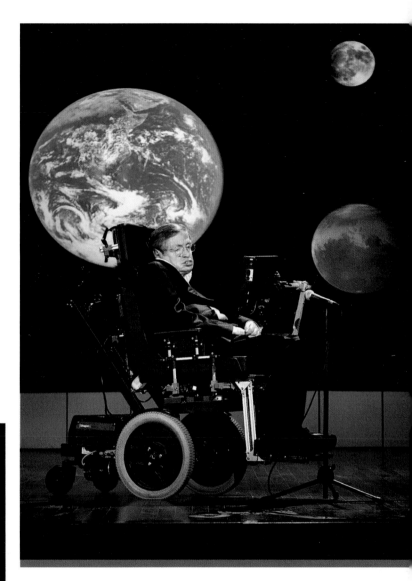

Left: Hawking delivers his speech for the Bulletin of the Atomic Scientists on 17 January 2007 in London.

Above: Hawking at the Opening Ceremony of the 2012 London Paralympics. He used his celebrity profile to warn of Doomsday threats.

Above: Hawking at a Bulletin of the Atomic Scientists meeting in 2007.

By the 2010s, Hawking's focus had shifted towards what he saw as two primary threats to humanity, outside of the usual suspects such as asteroid strike or nuclear war: rogue artificial intelligence (AI) and climate change (*see* box opposite). AI encompasses everything from computer programs that can make simple decisions on their own (such as distinguishing black from white) to the dream of sentient machines that will be at least as intelligent as humans. When Hawking discusses it, he is referring to the latter and, like other high-profile figures including the entrepreneur

and engineer Elon Musk, he has been vocal about the transformative and potentially catastrophic consequences of AI. In one article, for instance, he warned: "Success in creating AI would be the biggest event in human history. Unfortunately, it might also be the last, unless we learn how to avoid the risks.... Whereas the short-term impact of AI depends on who controls it, the long-term impact depends on whether it can be controlled at all."

In a forum on Reddit, the social news and web-content rating website, in 2015, Hawking expanded on these views, arguing that the

FIREBALL EARTH

The worsening threat of climate change seemed to convince Hawking to issue ever-diminishing timescales for the imminence of disaster. In 2016, for instance, he asserted that "a disaster to planet Earth… is a near certainty in the next thousand or 10 thousand years". At a summit in Beijing in 2017 he warned that within 600 years, runaway global warming could turn the planet into a fireball, citing a "worst-case scenario that Earth will become like its sister planet Venus, with a temperature of 250°C and raining sulphuric acid." Later that year Hawking slashed the countdown to just a century, claiming, in *Stephen Hawking: Expedition New Earth*, a BBC documentary to mark his seventy-fifth birthday, that "The human race only has 100 years before we need to colonize another planet."

Right: Artist's depiction of the surface of Venus. A fate that could await the Earth if Hawking's warnings are not heeded.

risk comes not from an evil AI but from an indifferent one. "You're probably not an evil ant-hater who steps on ants out of malice," he told one Reddit user, "but if you're in charge of a hydroelectric green energy project and there's an anthill in the region to be flooded, too bad for the ants. Let's not place humanity in the position of those ants." Hawking also warned that AI might worsen injustice and oppression, and was against AI weapons, saying he was anxious that "autonomous weapons will become the Kalashnikovs of tomorrow".

Right: Graph showing how average surface temperatures have increased over the last century, due to global warming.

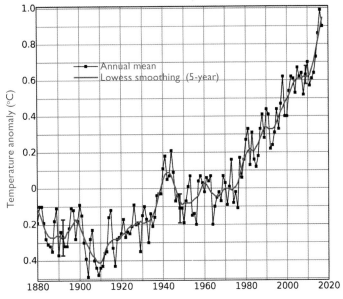

Global land–ocean temperature index

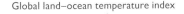

- Annual mean
- Lowess smoothing (5-year)

Temperature anomaly (°C)

M-THEORY

IN THE WAKE OF HIS SEMINAL WORK ON QUANTUM GRAVITY, BLACK HOLE ENTROPY AND HAWKING RADIATION, HAWKING HAD NURTURED THE AMBITION THAT HE COULD SUCCEED WHERE EINSTEIN HAD FAILED AND ACHIEVE A GREAT SYNTHESIS OF THE DIFFERENT AND INCOMPATIBLE THEORIES OF PHYSICS: A THEORY OF EVERYTHING (ToE – *SEE* PAGE 67).

Over the years Hawking's view on the possibility of such an achievement, and over the very existence of such a ToE, evolved and matured (*see* box opposite).

M-theory not only united the five different string theories, it also encompassed a complex theory of particle physics called 11-dimensional supergravity. M-theory seemed like an obvious candidate for the ToE, and indeed cosmologists such as Juan Maldacena were able to use it to create successful fusions of relativity and quantum physics for special cases. Hawking also embraced it, for example developing a mind-bending theory that our four-dimensional spacetime universe is a holographic projection of an 11-dimensional shadow membrane universe (*see* box page 126).

More generally, however, Hawking was moving away from the very idea of a ToE. By 2002 he was already suggesting that string theory, supergravity, M-theory and all the rest might be the only glimpses we can ever have of an ultimately unknowable bigger picture. He told an audience at the Paul Dirac Centennial Celebration, "Some people will be very disappointed if there is not an ultimate theory that can be formulated.... I used to belong to that camp, but I have changed my mind."

In Hawking's last major popular science book – his 2010 collaboration with American theoretical physicist Leonard Mlodinow, *The Grand Design* – Hawking described M-theory in detail. He also explored controversial

Above: American theoretical physicist Leonard Mlodinow, with whom Hawking collaborated towards the end of his life.

STRING REVOLUTIONS

Like most of the rest of the physics community, in the 1980s Hawking had joined in with the excitement at the development of string theory (see page 75). It seemed to be the most promising prospect for a ToE, especially when, in what was known as the "first string revolution", it was combined with the supersymmetry theory of particle physics (see page 66) to produce superstring theory.

However, further research on superstring theory revealed a complex picture, without a clear way forward. There were at least five different, self-consistent types of string theory, and it was not evident why any one of them should be more or less correct than the others. In 1995 there was a second string revolution, when physicists led by Edward Witten suggested that the competing 10-dimensional string theories could be reconciled into different aspects of a unified whole by adding an eleventh dimension. Adding another dimension to the strings effectively turned them into sheets or, in the terminology of the physicists, membranes, and accordingly this new approach was known as M-theory.

Below: Conceptualization of membranes in M-theory.

topics such as "model-dependent reality" (*see* box), argued that the universe was spontaneously created and thus there is no need for a creator god, and described his theories of top-down cosmology, multiple alternative realities and his version of the anthropic principle (*see* pages 128–31 for more on these).

THE SHADOW BRANE UNIVERSE

The classic conceptual analogy for the expanding universe is to picture spacetime as the surface of a balloon. As the balloon expands, spacetime grows bigger and everything in it moves away from everything else. A key aspect of this analogy is that it is limited: there is no inside to the balloon, and conceptualizing our universe as the skin of the balloon is purely analogous, as in reality it is four-dimensional, not two-dimensional. However, according to one of Hawking's interpretations of M-theory, this analogy should be extended much further. M-theory suggests that our 4-D universe is just one of many possible branes (short for membranes) – different configurations of the 11 dimensions the theory describes. There might be higher-dimensional branes beyond our own, not perceivable by us and hence sometimes called "shadow branes". Hawking suggested that our 4-D universe is only a holographic projection of such a higher-dimensional shadow brane, and that effectively our universe really is only the skin on the outside of a multi-dimensional balloon.

Above: Conceptual art showing the creation of a universe where two membranes collide.

Above: Conceptualization of a quantum gravity field. Describing quantum gravity was one of Hawking's grand ambitions.

MODEL-DEPENDENT REALISM

This is a theory which contends that we can only conceive of reality through our models or mental representations of it. Hawking and Mlodinow explain in their book: "There is no picture- or theory-independent concept of reality." Critics have pointed out that, in their book, Hawking and Mlodinow appear both philosophically ignorant – model-dependent realism is a re-hash of well-worn philosophical ideas going back to Immanuel Kant and beyond – and incoherent: they argue both that models create reality and that there are realities beyond the models which it is possible to approach more closely with better models.

TOP-DOWN COSMOLOGY

SINCE THE 1980S, HAWKING HAD STARTED TO HAVE RECOURSE TO THE ANTHROPIC PRINCIPLE (*SEE* BOX) TO UNDERPIN HIS IDEAS ON COSMOLOGY.

In 1999, he and Neil Turok suggested a model for the birth of the universe that would explain how it could have gone from the closed universe predicted by Hawking's no boundary proposal to one that is constantly expanding. They described a kind of primordial particle universe, which they likened to a pea since, although infinitesimally small, it would have had a similar mass to a pea, and would have been a small, slightly wrinkled sphere. They called this primordial pea an instaton (because it would have existed for just an instant), and used it to explain how an inflating universe resulted from the combination of gravity, matter, space and time that we observe in the universe today.

Above: Neil Turok with a statue of his friend Stephen Hawking, at the 2008 launch of the African Institute for Mathematical Sciences that he founded in Cape Town in South Africa.

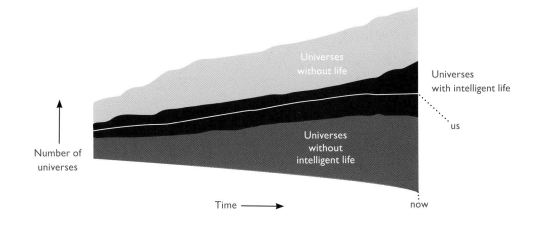

Number of universes

Universes without life

Universes with intelligent life

us

Universes without intelligent life

Time ⟶

now

Above: One explanation for the Goldilocks problem is that there are vast numbers of universes which support intelligent life, making it less remarkable that we should end up in one.

THE ANTHROPIC PRINCIPLE

Cosmologists sometimes refer to the Goldilocks principle in describing our universe: all of its fundamental parameters and natural laws are set up to be just right to allow the existence of intelligent life. This is a puzzle; it seems conceivable that any of these fundamental parameters, such as the strength of gravity or of the strong nuclear force, could be different from what they are, but if they were different, matter as we know it could not exist and intelligent life could never have existed. The settings of the universe seem to be arbitrarily precisely tailored (or "fine-tuned") for our benefit. One argument to explain this is to look at it the other way round: we are only able to observe the Goldilocks principle because the universe is indeed hospitable to life. If it were not, we would not be here and the question would not arise.

This is known as the anthropic principle. Hawking himself originally rejected the anthropic principle because he found it philosophically unsatisfying, arguing that it does not really explain anything. He came to reverse his opinion but many other cosmologists have not, arguing that the anthropic principle is either nothing more than circular logic or an appeal to some sort of guiding principle of universal design. An analogy might be with an experimental subject who correctly guessed the outcome of a coin toss a hundred times in a row. One could argue that this person was simply lucky, and that if they had not been so fortunate there would be nothing to explain. Alternatively, one could insist that there must be some meaningful explanation and seek to understand it.

A major flaw in the instanton theory, however, is that it predicts a myriad of possible evolutions for the subsequent universe, and almost none of them look anything like our own. To rescue the instanton, Hawking appealed to the anthropic principle. So long as one of the predicted outcomes allows the existence of intelligent life, it is possible to argue that the theory matches reality, because that is indeed the outcome that we observe. In 2006, Hawking collaborated with Belgian physicist Thomas Hertog to produce a theory

Above: Belgian physicist Thomas Hertog, Hawking's final collaborator.

that they called "top-down cosmology". This took as its starting point the quantum mechanical description of the universe that Hawking had developed for his no boundary proposal, and examined it through the lens of Feynman's sums over histories approach (*see* page 115). Together the two scientists argued that quantum mechanics describes not a single history, but an aggregate of all possible histories. All of these possibilities were contained within the initial waveform of the universe, in what quantum physicists describe as a "superposition". This is where multiple possible events coexist simultaneously until the superposition collapses to give a single outcome. Hawking and Hertog's idea was that, instead of starting with a set of initial conditions and working out which history must evolve from those – what might be termed a "bottom-up" approach – it makes more sense to start with the outcome and work backwards; hence the top-down approach. By looking at the eventual outcome, we can work out the selection process that turned an infinite number of initial possibilities into a single outcome. Hertog described this as the present "selecting" the past. And because the present universe is one that supports intelligent life, we can work backwards to show how the selection process went. Hawking had effectively moved from asking, as in his early days, "How did the universe begin?" to showing that this question is meaningless – the universe began in every way imaginable, and then evolved in the way that it was constrained to do by its eventual outcome.

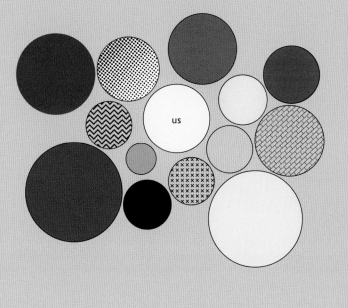

us

This is a new and ingenious use of the anthropic principle to reconcile one of the problems thrown up by M-theory, which is that the theory produces myriad, equally feasible predictions for how the universe could have evolved, almost none of which allow for the existence of intelligent life. Most M-theorists assume that an as-yet-unknown principle will allow them to "prune" away most of these possible universes, revealing underlying constraints that rule them out. But Hawking and Hertog were arguing that all of the different alternative universes really did exist, and that the top-down approach makes it possible to work out how we got to where we are now.

Left: Cosmic inflation could have led to the creation of multiple "bubble universes", each with different properties; top-down cosmology is an approach to explaining why we are in the one that supports intelligent life.

MANY START, ONE FINISHES

One way to explain the concept of top-down cosmology in relatively simple terms is to return to the analogy of the unfeasibly successful coin-tosser. The odds of achieving a hundred successful guesses in a row are 0.5^{100}, or roughly 1 in 10 million trillion trillion. Hawking and Hertog's approach to explaining the anthropic "fine-tuning" of the universe as the result of selection from multiple alternatives is analogous to starting with 10 million trillion trillion coin flippers and, with each flip, eliminating those who do not guess correctly. One would expect to end up with one person who had correctly guessed all 100 flips, without having to invoke ESP (extrasensory perception) or any other causal explanation. Similarly, even if there were many different ways that the universe could evolve from its Big Bang starting point, there is nothing mysterious about looking back from the vantage point of the one that did evolve.

CHAPTER 9

ON STAGE AND SCREEN

CULTURAL ICON

THROUGHOUT THE 1990S AND 2000S, STEPHEN HAWKING'S PROFILE ROSE EVER HIGHER. HE HAD LONG SINCE TRANSCENDED SCIENCE AND BECOME AN IMPORTANT PUBLIC FIGURE, BUT HE WAS NOW TRANSFORMING INTO SOMETHING GREATER — AN ARCHETYPAL FIGURE INVESTED WITH MEANING AND WISDOM FAR BEYOND THE REALITY OF THE PERSON.

As both a driver and consequence of this transformation, Hawking was becoming a cultural icon, immediately recognizable from his profile and distinctive voice, on a par with figures such as Darth Vader or Marilyn Monroe.

He had featured in documentaries since the 1970s, but now he became a fictional presence

Above: Hawking alongside Einstein, Newton and the android Data, in an episode of *Star Trek: The Next Generation*.

as well. In 1992 he was a key character in *The Voyage*, an opera by Philip Glass, who had written the music for Errol Morris's film of *Brief History*. In 1993 Hawking met Leonard Nimoy at an event; when Nimoy learned that the scientist was a *Star Trek* fan he arranged for him to play a memorable guest-star role in an episode of *Star Trek: The Next Generation*. Hawking appears alongside Newton and Einstein in a virtual poker game with one of the show's main characters, and gets to display his humour and winning smile. It was telling that he did not demur from being ranked alongside such elevated company. That same year Hawking's voice featured on a Pink Floyd music track.

Alongside his *Star Trek* appearance, one of Hawking's other pop-culture touchstones was his recurring role in *The Simpsons*, the long-running US animated series. In 1999 he flew to Los Angeles to voice a part in which he discussed a doughnut-shaped universe with Homer, a guest appearance that spawned a Stephen Hawking *Simpsons* action figure. Hawking would appear in the series again in 2005, and in its sister series *Futurama*, and he would go on to feature repeatedly in the sitcom *The Big Bang Theory* in the 2010s. He also inspired a 1999 play by Robin Hawdon, called *God and Stephen Hawking*.

Right: Hawking at the premiere of the film of *A Brief History of Time* at the Samuel Goldwyn Theater in Los Angeles.

Below: Scene from *The Simpsons* episode "They Saved Lisa's Brain", one of several to feature a cameo by a cartoon version of Hawking.

MASTER OF DOCUMENTARIES

Multiple appearances on television across the world helped to make Hawking the most recognizable and famous scientist on the planet, a status on which documentary makers were keen to cash in. His name was attached to at least eight different documentaries or documentary series, including *Stephen Hawking's Universe* (1997), *Stephen Hawking: Master of the Universe* (2008), *Into the Universe with Stephen Hawking* (2010–11) and *Genius by Stephen Hawking* (2016).

Right: Hawking with other noted brainiacs Arthur C. Clarke and Magnus Magnusson, in the 1988 documentary *Masters of the Universe*.

HAWKING THE AUTHOR

IN ADDITION TO NUMEROUS ACADEMIC PAPERS AND SOME ACADEMIC BOOKS, HAWKING WROTE OR CO-WROTE 12 POPULAR SCIENCE BOOKS. NONE WOULD REPEAT THE RECORD-SMASHING SUCCESS OF HIS FIRST, *A BRIEF HISTORY OF TIME* (1988), BUT MANY WOULD BECOME BESTSELLERS IN THEIR OWN RIGHT.

Hawking's initial follow up to *Brief History* was a collection of essays – *Black Holes and Baby Universes and Other Essays* (1993) – which collected texts on the cosmology of black holes and how they might give rise to new universes, along with more personal material. *The Nature of Space and Time* (1996) chronicled a scientific debate between Hawking and Roger Penrose, on physics and the philosophy of physics. *The Universe in a Nutshell* (2001) was intended as a follow-up to *Brief History*. It started life as a collection of essays, but with help from editor (and his later biographer) Kitty Ferguson, Hawking turned it into a wide-ranging account of the current state of theoretical cosmology, delving into the arcane world of string theory, M-theory, supersymmetry, inflation, bubble universes, quantum cosmology and sums over histories. Heavily illustrated, it was an ambitious attempt to achieve the same kind of challenging discourse on cutting-edge cosmology that *Brief History* had managed to sell to millions, and in 2002 it won the Royal Society's prestigious Aventis Prize for Science Books.

NEVER SAY NEVER

The runaway success of *Brief History* prompted inevitable questions (presumably from his publishers, among others) about whether Hawking would pen a sequel. At the time he was quite definite that he would not, sarcastically querying what it might be titled. "What would I call it? *A Longer History of Time? Beyond the End of Time? Son of Time?*" The answer, it turned out, was *A Briefer History of Time*, the 2005 follow-up he co-wrote with Leonard Mlodinow. This expanded and clarified on some of the topics from the original book, updating it to take account of recent developments.

Above: Kitty Ferguson, who worked with Stephen Hawking on *The Universe in a Nutshell* and later wrote his biography.

CHILDREN'S BOOKS

Starting in 2007, Stephen and his daughter Lucy (a journalist and novelist) co-wrote the first in what would become a series, *George's Secret Key to the Universe*. The series follows the adventures of two kid cosmologists, George and Annie, while trying to convey a range of science and cosmology lessons. The fifth instalment in the series, *George and the Blue Moon*, was published in 2016.

Left: Lucy Hawking with the first book she co-authored with her father, *George's Secret Key to the Universe*.

Hawking quickly followed this up with *On the Shoulders of Giants: The Great Works of Physics and Astronomy* (2002), a curated collection of classic texts from great figures in the history of science and cosmology, from Copernicus to Einstein. Hawking provided commentary, mini-biographies and accompanying critical essays. In 2005 he published a companion volume about mathematics, *God Created the Integers: The Mathematical Breakthroughs That Changed History,* which included extracts from 31 of the most important works in the history of mathematics. A third work in the same vein, treating quantum mechanics, *The Dreams That Stuff Is Made Of: The Most Astounding Papers on Quantum Physics and How They Shook the Scientific World*, was published in 2011. His final book as solo author was his autobiography, *My Brief History* (2013).

Above: Members of the Stephen Hawking fan club at the Gold Star Sardine bar, Chicago, with club co-founder Susan Anderson in the foreground.

DANCING WITH THE STARS

PART OF STEPHEN AND JANE HAWKING'S STORY HAD ALREADY BEEN
FILMED, AS *HAWKING*, IN 2004 (*SEE* PAGE 141), BUT IN 2014 A NEW BIOPIC
CATAPULTED JANE TO GLOBAL FAME WHILE CEMENTING STEPHEN'S PLACE AS
THE MOST FAMOUS SCIENTIST OF HIS ERA.

Above: Eddie Redmayne as young Stephen Hawking in the 2014 hit film *The Theory of Everything*. Redmayne's challenge was to portray Hawking over a nearly 30-year span, during which his illness manifested and progressed.

The Oscar-winning film *The Theory of Everything* was based on Jane's memoir, originally published in the UK as *Music to Move the Stars*, but later re-issued as *Travelling to Infinity*. The book had caught the attention of screenwriter Anthony McCarten back in 2004, to the extent that he began to adapt it into a screenplay long before persuading Jane to grant him the rights. McCarten, after 2009 in tandem with his producing partner Lisa Bruce, spent around eight years convincing Jane to agree to the adaptation. Bruce recalled that it took "a lot of conversation, many glasses of sherry, and many pots of tea".

For the lead role of Stephen, several actors were considered over the long gestation of the movie, including Andrew Garfield and Michael Fassbender. However, after the director James Marsh met with British actor Eddie Redmayne in a London pub, he was offered the role without even needing to audition. British actress Felicity Jones was cast as Jane.

The most anxiety-provoking moment in his preparation was meeting Hawking himself. Redmayne rambled on, filling the inevitable silence with inane chatter about star signs that prompted Hawking to type out a rather acid

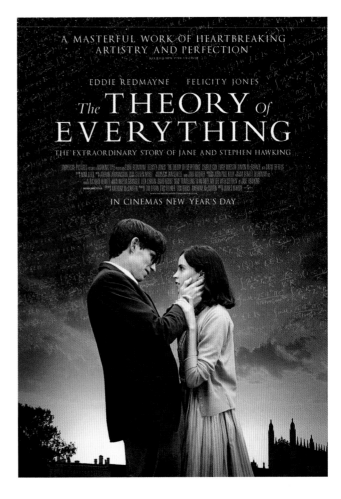

Above: The poster for the film *The Theory of Everything*, featuring romance, passion and the obligatory equations on a blackboard.

PREPARING TO ACT WITH ALS

Redmayne recalled seeing Hawking's iconic silhouette crossing the road while he was a student at Cambridge, but now he was faced with the reality of trying to portray every stage in the progression of a debilitating disease that produces profound bodily changes. He spent four months reading and watching everything he could about Hawking, although he admitted that little of the science made any sense to him. To capture the physicality of the role, Redmayne worked with choreographer Alexandra Reynolds, and regularly visited a neurology clinic to meet more than 30 ALS patients.

Below: Redmayne, brandishing the BAFTA award he has just won, with Hawking at the BAFTA British Academy Film Awards at the Royal Opera House in 2015.

observation: "I'm an astronomer, not an astrologer." During filming Hawking visited the set while they were recreating the May Ball (*see* page 26), drinking champagne from a teaspoon, and when the film was complete he came to a preview screening. Cast and crew were relieved that his reaction was positive. He called the film "broadly true" and let them use his copyrighted trademark synthesized voice in place of the fake substitute they had been making do with. Later he emailed the director to admit that "there were certain points when he thought he was watching himself".

The movie proved to be a critical and commercial success, with Redmayne in particular winning spectacular reviews for his portrayal of Hawking. Redmayne himself spoke of how Hawking's friends and family had pointed out Stephen's remarkably expressive eyebrows, and he made good use of his own in producing his memorable performance. Stephen and Jane Hawking accompanied Redmayne and Jones on the red carpet at the British premiere of the film.

Awards season would bring glory to the filmmakers and raise the profile of the Hawkings even higher. At the British Academy Film Awards (BAFTAs), the film received 10 nominations and went home with Outstanding British Film, Best Leading Actor for Redmayne and Best Adapted Screenplay for McCarten. Redmayne went on to win a Golden Globe and the Screen Actors Guild award, and was a red-hot favourite for the Best Actor Oscar. At the 2015 Academy Awards he duly won, dedicating the prize to "all of those people around the world battling [motor neurone disease]", and saying that it really belonged to Stephen, Jane and their "exceptional family". The contrasting verdicts on the film from Jane and Stephen were extremely telling: she felt that it had perhaps too much science and too little emotion, while he thought the opposite.

HAWKING VERSUS *THE THEORY OF EVERYTHING*

It is instructive to compare the two film versions of Jane and Stephen's story. *The Theory of Everything* covers a much broader span of Hawking's life, and includes his near-death crisis and tracheotomy, along with much more of the couple's personal lives, including Jane's relationship with Jonathan, and Stephen leaving her for Elaine. However, central to both movies are Jane and Stephen's initial meetings, his diagnosis and the breakthrough work in his PhD on singularities, after meeting Roger Penrose. *Hawking* is far more accurate and true to life, whereas *Theory of Everything* takes very broad liberties with timeline and characters – for example, combining Hawking's college years, in order to locate in Cambridge incidents that actually took place in Oxford. While neither film has a great deal of scientific content, *The Theory of Everything* is notably superficial in its treatment of Hawking's theories and science in general, complete with a hackneyed sequence in which he writes out esoteric equations on a blackboard. Any objective judgement of the film must concur with Stephen's own assessment that it privileges emotion over science.

Right: Redmayne and Felicity Jones, as Jane, re-creating a scene from the courtship of the Hawkings in the early 1960s.

CHAPTER 10

LEGACY

HAWKING'S LAST YEARS

STEPHEN HAWKING CONTINUED TO WORK UP UNTIL THE VERY END OF HIS LIFE, BOTH AT THE COAL FACE OF COSMOLOGY AND AT HIS BROADER MISSION OF SCIENCE COMMUNICATION AND PUBLIC ENGAGEMENT.

Alongside headlining the Breakthrough Initiatives relating to space exploration (*see* "Reaching for the Stars", page 146), Hawking delivered the BBC Reith Lectures for 2016. This prestigious lecture series gives a platform to a public intellectual, and Hawking used it to introduce his latest work on black holes, which saw him collaborating with others to offer a bold but fiendishly arcane solution to

the black hole information paradox he had created back in 1981.

This new approach, conceived in conjunction with Harvard physicist Andrew Strominger and his own Cambridge colleague Malcolm Perry, claimed that black holes preserve information because they have "soft hair". This is a reference to John Wheeler's dictum that black holes have no hair, which is to say that there are almost

Above: Hawking in June 2015 in his office at Cambridge University. He continued working up to the final weeks of his life.

Right: Andrew Strominger of the Harvard University Department of Physics, with whom Hawking collaborated on a new theory about the black hole information paradox.

no variable features that can distinguish one black hole from another. The no-hair theorem had been one of the key elements of the black hole information paradox (*see* pages 68–71). Hawking, Strominger and Perry now proposed that in fact the empty space around the event horizon preserves traces of radiation that has fallen into the black hole in the form of "soft particles" – photons and gravitons (particles that mediate the force of gravity) that have zero energy but do have other distinctive characteristics. In other words, Hawking was arguing, he had been wrong to assume that black holes have no hair: in fact they have soft hair. This soft hair, Hawking et al now proposed, preserves black hole information even when the black hole has evaporated.

The precise mechanics of this preservation were unclear and there was pushback from the cosmology community, but the soft hair proposal did make testable predictions. For instance, it raised the intriguing possibility that these soft particles leave a kind of "gravitational memory", a signature in the fabric of spacetime, which might be detected (or perhaps reconstructed) by comparing measuring devices (such as clocks, rulers or their hi-tech equivalents) that have changed minutely in response to these incredibly subtle alterations in spacetime.

In 2016 Hawking lent his likeness to the computer game *Science Kombat*, in which famous scientists from history battle using characteristic special powers (Hawking's was black hole attack). He also campaigned against Brexit, continuing his lifelong history of espousing left-wing and liberal causes (*see* box).

HAWKING'S POLITICS

Hawking was a long-time supporter of the Labour Party, on the left wing of British politics, and was a passionate advocate for related causes, especially the National Health Service (NHS), Britain's universal health system. He frequently found himself at odds with government spending priorities, and was said to have turned down a knighthood in the late 1990s because he was so angry about government failure to fund British science adequately. In the last year of his life he embarked on what the BBC described as "an epic feud" with the minister then in charge of the NHS, Jeremy Hunt.

Below: Then British Health Secretary Jeremy Hunt, with whom Hawking publicly disputed over funding for Britain's National Health Service.

REACHING FOR THE STARS

TOWARDS THE END OF HIS LIFE, HAWKING PARLAYED HIS STATUS AND PROFILE INTO BECOMING THE DRIVING FORCE BEHIND A RANGE OF AMBITIOUS AND EXPENSIVE INITIATIVES. WITH EXTRAORDINARY ENERGY AND DYNAMISM, HE FOUNDED A NEW CENTRE FOR CUTTING-EDGE RESEARCH IN CAMBRIDGE, AND HEADLINED A VISIONARY PROGRAMME TO EXPLORE THE COSMOS IN SEARCH OF INTELLIGENT LIFE.

Hawking had spent most of his career working at the Department of Applied Mathematics and Theoretical Physics (DAMTP) at Cambridge. In 2007 he headlined the establishment of a department, the Centre for Theoretical Cosmology (CTC), as part of Cambridge's Centre for Mathematical Sciences, a hi-tech, modern office complex. The CTC incorporated the DAMTP's research group on cosmology, and was also home to the COSMOS supercomputer. In 1997 Hawking had pulled together the consortium that set up COSMOS

BREAKTHROUGH STARSHOT

A futuristic initiative to develop a radical new form of spaceship capable of traversing interstellar distances within a human lifetime, Breakthrough Starshot was announced with much fanfare by Hawking and Milner in 2016. Its aim is to develop nano-ships: very light miniature spacecraft – essentially satellites on a chip – weighing no more than four grams. These will be attached to solar sails and accelerated to around a fifth of the speed of light by very high-power lasers. A solar sail is a surface on which photons (from the sun or an artificial source) can impact, transferring their minuscule momentum to accelerate it in similar fashion to a sail on Earth.

Most spaceship designs involve relatively massive spacecraft. Simple physics shows that, even with colossal energy expenditure far outstripping their fuel-carrying capabilities, such spaceships can only approach speeds that make interstellar journeys unfeasible. The closest star system to Earth, Alpha Centauri, is 40 trillion km (25 trillion miles, 4.37 light years) away. The fastest spacecraft yet created would take 30,000 years to get there. Breakthrough Starshot aims to boost their nano-ships to previously undreamed-of speeds. As Hawking boasted at the launch, "Such a system could reach Mars in less than an hour, or reach Pluto in days, pass Voyager in under a week and reach Alpha Centauri in just over 20 years."

Above: A solar sail, one of the technologies intended to help accelerate Starshot probes to relativistic speeds.

Opposite: Hawking at the Centre for Mathematical Sciences, the hi-tech modern development at which he was based from 2007.

Below: Hawking at the Breakthrough Starshot launch in 2016, alongside other luminaries including Freeman Dyson, Ann Druyan, Avi Loeb, Mae Jemison and Pete Worden.

as a supercomputer facility dedicated to research in cosmology, astrophysics and particle physics. In 2012 an upgraded version of COSMOS came on line, providing the largest shared-memory computer in Europe.

A more headline-grabbing initiative came with Hawking's involvement in the Breakthrough Initiatives, a series of ambitious space exploration projects bankrolled by internet investor Yuri Milner. Hawking provided the public face for the launch of projects such as Breakthrough Listen (a $100 million astronomy programme designed to look for evidence of intelligent life beyond Earth), Breakthrough Watch (a project to design new methods to observe extra-solar planets capable of supporting life) and Breakthrough Starshot (*see* box).

LAST WORDS

ON WEDNESDAY, 14 MARCH 2018, STEPHEN HAWKING'S CHILDREN LUCY, ROBERT AND TIM RELEASED A JOINT STATEMENT ANNOUNCING THAT THEIR FATHER HAD PASSED AWAY EARLY THAT MORNING AT HIS HOME IN CAMBRIDGE.

They paid tribute to his "work and legacy... his courage and persistence... [and] his brilliance and humour", and quoted a saying of his: "'It would not be much of a universe if it wasn't home to the people you love.' We will miss him forever."

No precise cause of death was announced, and most sources simply put it down to complications due to Hawking's ALS. The progressive degeneration of breathing and swallowing muscles makes ALS sufferers extremely prone to pneumonia (lung infections), and, given that Hawking's life had been threatened by bouts of pneumonia on multiple occasions, it seems likely that this was the cause of his death.

Tributes flooded in from around the world and the incredible range of spheres of life represented amongst them bore testament to the breadth of Hawking's appeal and the magnitude of his stature. In addition to figures from science and academia, those paying tribute included

politicians and religious figures, entertainers and movie stars, and, of course, friends and family.

Hawking's funeral was held on 31 March 2018 at the university church, Great St Mary's, in Cambridge. Crowds of well-wishers braved poor weather to stand outside the church, while the congregation heard speakers including actor Eddie Redmayne, Stephen's eldest son Robert, former student Prof Fay Dowker and Astronomer Royal Martin Rees. The church bell rang 76 times – once for each year of Hawking's life – and his coffin was decorated with white lilies, to represent the universe, and white roses to represent the polar star.

Opposite: The interment of Stephen Hawking's ashes at Westminster Abbey in London on 15 June 2018. His daughter Lucy places flowers, as her mother (second from left) looks on.

Above: Crowds of well-wishers brave the rain to watch the arrival of the funeral cortege of Stephen Hawking at the University Church of St Mary the Great in Cambridge, 31 March 2018.

BETWEEN THE SHOULDERS OF GIANTS

On 15 June 2018, Hawking's ashes were interred at Westminster Abbey, between the remains of Sir Isaac Newton and Charles Darwin. According to his wishes, his memorial stone was engraved with what he – and the rest of the scientific world – regarded as his greatest achievement: the equation for the entropy of a black hole, also known as the Bekenstein-Hawking entropy equation (see pages 52–59 for more on this). Hawking evidently felt that this equation was the best possible epitaph for his extraordinary life and work.

The Bekenstein-Hawking equation:

$$S = \frac{\pi A k c^3}{2hG}$$

THE POSTHUMOUS PAPER

EVEN AFTER HIS DEATH, STEPHEN HAWKING WAS NOT FINISHED WITH STIRRING UP THE WORLD OF COSMOLOGY. HE HAD CONTINUED WORKING UP UNTIL THE VERY END OF HIS LIFE, SUBMITTING A PAPER FOR PUBLICATION ON 4 MARCH 2018, LESS THAN TWO WEEKS BEFORE HIS DEATH. THAT PAPER, CO-AUTHORED WITH BELGIAN PHYSICIST AND LONG-TIME COLLABORATOR THOMAS HERTOG, WAS TITLED "A SMOOTH EXIT FROM ETERNAL INFLATION?".

Above: Conceptualization of the multiverse hypothesis, showing many different universes existing in parallel.

The fact that this was Hawking's last work inspired a lot of hype, with claims that it was his greatest work or that it would have won him a Nobel prize. In practice, it was only an initial exploration of a very simplified mathematical model that the authors themselves called a "toy model". The paper tackles one of the problems thrown up by the theory of cosmological inflation – "eternal inflation" (*see* box). Eternal inflation suggests that our universe is just one of an effectively infinite number of bubble universes or multiverses, each of which could have its own set of physical parameters for the laws of nature (for example, a different constant for the speed of light). Physicists dislike this kind of infinite choice, because it explains nothing. How can we account for the nature of our universe, when there an infinite number of alternative natures? "The traditional formulation of the multiverse theory could not be falsified," explains Avi Loeb, chair of the Harvard Astronomy Department, "if anything that can happen will happen an infinite number of times."

Hawking and Hertog's paper develops a mathematical argument that inflation is constrained, so that when a region of spacetime stops inflating ("exits from inflation") it does so in a smooth fashion so that none of the resulting bubble universe will undergo further inflation. This work forms part of Hawking's ongoing attempt to limit the proliferation of multiverses and demonstrate that those bubble universes that do form must be constrained to be like the one we actually observe. In other words, right up until the end, he was attempting to show that it is possible to address the question of where the laws of physics come from, and that it might yet be possible to get a definitive answer that explains the universe.

ETERNAL INFLATION

Cosmological inflation is a theory that was created to explain how our universe came to look, to theoretical physicists, relatively smooth and flat. It proposes that immediately after the birth of the universe a mysterious kind of energy caused the fabric of spacetime itself to inflate or expand. But a problem with the mathematics of this inflation is that they predict that when more spacetime is created in this way, at least some portions of it will then undergo their own inflation. Every time this happens a new bubble universe is created, and even if most of regions of these universes stop inflating, there will always be at least a few regions that do not stop. In this way inflation keeps going forever (hence "eternal inflation") and the result must be an infinite number of bubble universes or multiverses.

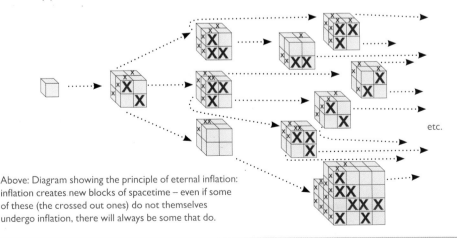

etc.

Above: Diagram showing the principle of eternal inflation: inflation creates new blocks of spacetime – even if some of these (the crossed out ones) do not themselves undergo inflation, there will always be some that do.

HAWKING'S SCIENTIFIC LEGACY

IT IS DIFFICULT TO DISENTANGLE THE SCIENTIFIC LEGACY OF STEPHEN HAWKING FROM THE WIDER PICTURE OF HIS CULTURAL IMPACT.

Many of his death notices and obituaries, for instance, refer to him as having been the smartest man alive, the greatest scientist of his era, the greatest cosmologist since Einstein, and so on.

Such hyperbolic praise inevitably provokes a backlash, and much has been made of the less worshipful attitude of Hawking's peers.

For instance, a survey of physicists around the millennium by *Physics World* magazine, which asked them to name the most important living physicists, did not place Hawking near the top 10. So it is essential to separate the public profile of Hawking, the icon of genius, from a more sober assessment of his scientific impact.

Above: Hawking and other stars at the launch of his eponymous medal for science communication at the Starmus festival in 2015.

Right: Another of Hawking's legacies: a school in East London named after him, for children with learning difficulties.

Like many ground-breaking physicists, Hawking did his best work in his youth. His PhD work on the cosmological singularity helped to set the terms of debate for Big Bang studies thereafter. It is, however, his subsequent work on black holes, specifically his discovery of black hole entropy and Hawking radiation, and the equation that describes them, which is generally regarded as his major achievement and contribution to science (see box below). This is the work that will stand the test of time as his greatest scientific legacy.

WHY IS HAWKING'S EQUATION SO IMPORTANT?

Hawking's work on the cosmological singularity was a brilliant exploration of general relativity, the theory about space, time and the large-scale universe. In his work on black hole entropy, Hawking accomplished the incredible achievement of uniting gravity and other physical forces with quantum mechanics, and his equation demonstrates how. The terms it contains include: Newton's constant, relating to gravity; Planck's constant, relating to quantum mechanics; the speed of light, which relates to Einstein's relativity; and the Boltzmann constant, which relates to thermodynamics. Thus it reveals a profound relationship between gravity, quantum mechanics, relativity and thermodynamics, and demonstrates that black holes are the phenomenon that is key to exploring this relationship.

WHY NO NOBEL?

Hawking often joked about receiving a Nobel prize, and many of his fans must be mystified as to how one eluded him. The reason is that the Nobel prize is almost always given out for proof of a theory, rather than for the theory itself. So, for instance, Peter Higgs had to wait for the actual discovery of the Higgs boson before he won his Nobel for predicting its existence. Hawking's work was theoretical, and dealt with such extreme phenomena that evidential proof of the theories has not yet been discovered. Hawking radiation, for instance, could only be detected from subatomic-scale black holes, and these have yet to be observed or created. Similarly some of his work on inflation and the Big Bang made predictions about primordial gravitational waves, but again these have yet to be detected. Such evidence may well arrive, perhaps even in the near future, but Nobel prizes are not awarded posthumously, so Hawking will never receive one.

Below: Simulation of micro black hole formation, an achievement that would have won a Nobel prize for Hawking.

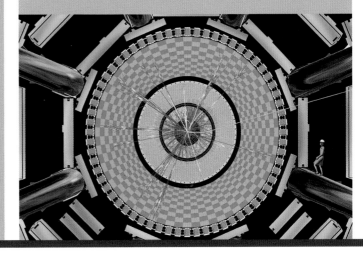

GLOSSARY

Absolute Zero: Temperature equivalent to -273.15°C (or 0 on the absolute Kelvin scale); it is the lowest possible temperature, at which atoms stop vibrating.

Accretion disc: Dust and gas orbiting a central body such as a young star or black hole, squeezed into a disc by gravity; forces acting on the disc can cause it to emit electromagnetic radiation.

Anthropic principle: The universe is as it is because intelligent (human) life is here to observe it; cosmological theories must allow for such life.

Atom: Basic building block of all normal matter, consisting of a nucleus (which is itself composed of positively charged protons and zero-charged neutrons) orbited by a cloud of negatively charged electrons.

Big Bang: The cosmic expansion or explosion from a single point, which is thought to have created our universe 13.7 billion years ago. Time and space were created in the Big Bang, so nothing can have come before or outside it.

Black hole: A bottomless pit in spacetime, from which no light or matter can escape, caused when a massive body collapses under its own gravity into a point of infinite density called a singularity.

Boson: Energy-carrying particle.

Conservation: Principle that quantities such as matter, energy or information cannot appear out of or disappear into nothing, so that the total amount must be conserved before and after any change or interaction.

Cosmic background microwave radiation: Low temperature microwave radiation diffused throughout the universe; remnant heat of the Big Bang, known as the "afterglow of creation".

Cosmology: The science of the origin and development of the universe.

Electromagnetism: Fundamental force exerted by an electromagnetic field on charged particles; waves in this field are known as radiation. Mediated by exchange of photons.

Electron: Fundamental subatomic particle with single negative charge, usually found orbiting the nucleus of an atom.

Energy: Ability to do work or cause change.

Entropy: Disorder, randomness, unavailable heat energy – the degree of disorder or uncertainty in a system. The second law of thermodynamics is that entropy always increases over time.

Equivalence: Of mass and energy – the fact that mass is concentrated energy, akin to energy in a different phase, as described by the equation $E=mc^2$. Any mass thus has an associated energy and any energy has an associated mass.

Event horizon: Region of space around a singularity where force of gravity exceeds escape velocity of light, so that anything that crosses this threshold cannot escape.

Force: Interaction between two objects such that energy is transferred between them; mediated by exchange of particle-like packets of energy known as gauge bosons.

Gravitational wave: A wave carrying energy in a gravitational field; ripples in the fabric of spacetime.

Graviton: Hypothetical elementary particle that mediates the force of gravity.

Gravity: Fundamental force; attraction between any two massive bodies.

Inflation: The hypothesis that almost immediately after the Big Bang time and space expanded much faster than the speed of light, doubling in size at least 90 times in less than 10^{-32} of a second.

Light year: Distance travelled by light in a year – 9.46 trillion km (5.9 trillion miles).

Mass: Amount of matter in a body.

Matter: Anything that has mass and volume.

Multiverse: Multiple possible universes that diverged (and may constantly be diverging) from our own, and which may have different laws of physics.

Neutron: Subatomic particle with no electric charge.

Nucleosynthesis: The process by which atomic nuclei are synthesized, as protons and neutrons fuse together.

Nucleus: The central part of an atom, where almost all the mass is concentrated, comprised of protons and neutrons.

Paradox: Scenario or system leading to logically coherent but incompatible outcomes.

Photon: Packet of electromagnetic energy; particle of electromagnetic radiation or light. Boson mediating electromagnetic force.

Planck constant: Units of measurement devised to simplify mathematics of quantum physics, which describe scales at the quantum level, including the smallest possible units of time and space and the highest possible temperatures and energies.

Proton: Positively charged subatomic particle in the nucleus of atoms.

Quantum: The smallest packet or chunk into which something can be divided; description of the universe at the smallest scale, in which everything is divided into discrete units.

Quantum mechanics/physics/theory: Description of the universe at very small scales, at which objects, particles or bodies can be considered in isolation from their surroundings, and behave in probabilistic and not deterministic fashion, as both waves and particles.

Redshift: Decrease in frequency and increase in wavelength of light waves from a source receding relative to an observer, causing shift to the red end of the spectrum.

Relativity: Theory describing how space and time are relative to one another, not absolute, and how they relate to matter and energy.

Singularity: Point in space where the curvature of spacetime, or the density of matter, becomes infinite, and where laws of physics break down.

Spacetime: Conception of the universe in which time is considered as a fourth dimension alongside the three spatial ones, and all four together make up a single fabric or continuum.

Steady state: Model of the universe with no beginning or end, where new matter and energy are constantly created to replace matter and energy that disappear.

Topology: Study of geometry and spatial properties that do not depend on shape changes such as bending or stretching.

Wormhole: Hypothetical tunnel or bridge between two non-adjacent regions of spacetime, where two singularities have joined together.

INDEX

PICTURE CREDITS